Ben Stacy Jerrik (Ed.)

# Nature Worship

Ben Stacy Jerrik (Ed.)

# Nature Worship

Nature

Part Press

**Imprint**



Publisher:
Part Press is a trademark of
International Book Market Service Ltd., 17 Rue Meldrum, Beau Bassin, 1713-01 Mauritius
Email: info@bookmarketservice.com
Website: www.bookmarketservice.com

Published in 2012

Printed in: U.S.A., U.K., Germany. This book was not produced in Mauritius.

**ISBN: 978-613-8-63157-6**

# Contents

## Articles



## References

# Nature worship

**Nature worship** describes a variety of religious, spiritual and devotional practices that focus on natural phenomenon. A nature deity can be in charge of nature, the biosphere, the cosmos or the universe. Nature worship can be found in panentheism, pantheism, deism, polytheism, animism, totemism, shamanism and paganism where deities are viewed as the embodiment of natural forces. Common to most forms of nature worship is a spiritual focus on the individual's connection to the natural world and reverence towards it.

## Forms of nature worship

- Fire worship
- Tree worship
- Animal worship
- Star worship
- Sacred mountains
- Sacred groves
- Sacred herbs
- Holy well
- Megalith
- Standing stone
- Stone circle
- Thunder god
- Totem
- Sky deity
- Water deity
- Naturalistic pantheism
- Naturalistic spirituality
- Gaia philosophy

## See also

- The sacred way
- Folk religion
- Pantheism
- Shamanism
- Earth religion
- Neopaganism
- Goddess worship (disambiguation)
- Natural religion (disambiguation)
- Wildlife totemization
- Category:Sacred sites

# Nature

**Nature**, in the broadest sense, is equivalent to the **natural world**, **physical world**, or **material world**. "Nature" refers to the phenomena of the physical world, and also to life in general. It ranges in scale from the subatomic to the cosmic.

Hopetoun Falls, Australia

The word *nature* is derived from the Latin word *natura*, or "essential qualities, innate disposition", and in ancient times, literally meant "birth".[1] *Natura* was a Latin translation of the Greek word *physis* (φύσις), which originally related to the intrinsic characteristics that plants, animals, and other features of the world develop of their own accord.[2] [3] The concept of nature as a whole, the physical universe, is one of several expansions of the original notion; it began with certain core applications of the word φύσις by pre-Socratic philosophers, and has steadily gained currency ever since. This usage was confirmed during the advent of modern scientific method in the last several centuries.[4] [5]

Bachalpsee in the Swiss Alps

Within the various uses of the word today, "nature" often refers to geology and wildlife. Nature may refer to the general realm of various types of living plants and animals, and in some cases to the processes associated with inanimate objects – the way that particular types of things exist and change of their own accord, such as the weather and geology of the Earth, and the matter and energy of which all these things are composed. It is often taken to mean the "natural environment" or wilderness–wild animals, rocks, forest, beaches, and in general those things that have not been substantially altered by human intervention, or which persist despite human intervention. For example, manufactured objects and

Lightning strikes during the eruption of the huge Galunggung volcano, West Java, in 1982.

human interaction generally are not considered part of nature, unless qualified as, for example, "human nature" or "the whole of nature". This more traditional concept of natural things which can still be found today implies a distinction between the natural and the artificial, with the artificial being understood as that which has been brought into being by a human consciousness or a human mind. Depending on the particular context, the term "natural" might also be distinguished from the unnatural, the supernatural, or synthetic.

## Earth

Earth (or, "the earth") is the only planet presently known to support life, and its natural features are the subject of many fields of scientific research. Within the solar system, it is third nearest to the sun; it is the largest terrestrial planet and the fifth largest overall. Its most prominent climatic features are its two large polar regions, two relatively narrow temperate zones, and a wide equatorial tropical to subtropical region.[6] Precipitation varies widely with location, from several metres of water per year to less than a millimetre. 71 percent of the Earth's surface is covered by salt-water oceans. The remainder consists of continents and islands, with most of the inhabited land in the Northern Hemisphere.

View of the Earth, taken in 1972 by the Apollo 17 astronaut crew. This image is the only photograph of its kind to date, showing a fully sunlit hemisphere of the Earth.

Earth has evolved through geological and biological processes that have left traces of the original conditions. The outer surface is divided into several gradually migrating tectonic plates. The interior remains active, with a thick layer of plastic mantle and an iron-filled core that generates a magnetic field.

The atmospheric conditions have been significantly altered from the original conditions by the presence of life-forms,[7] which create an ecological balance that stabilizes the surface conditions. Despite the wide regional variations in climate by latitude and other geographic factors, the long-term average global climate is quite stable during interglacial periods,[8] and variations of a degree or two of average global temperature have historically had major effects on the ecological balance, and on the actual geography of the Earth.[9] [10]

### Geology

Geology is the science and study of the solid and liquid matter that constitutes the Earth. The field of geology encompasses the study of the composition, structure, physical properties, dynamics, and history of Earth materials, and the processes by which they are formed, moved, and changed. The field is a major academic discipline, and is also important for mineral and hydrocarbon extraction, knowledge about and mitigation of natural hazards, some Geotechnical engineering fields, and understanding past climates and environments.

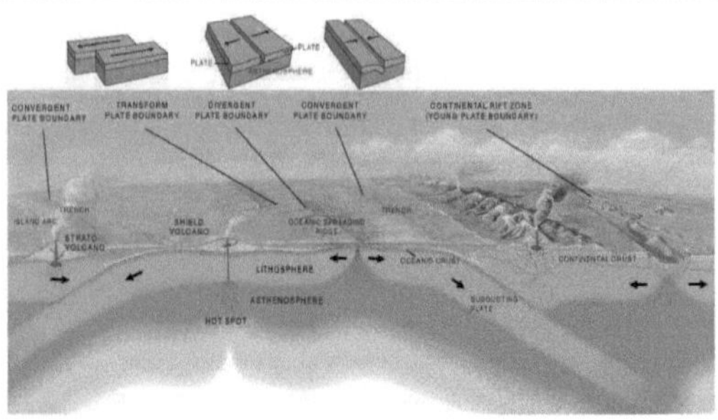


Three types of geological plate tectonic boundaries.

### Geological evolution

The geology of an area evolves through time as rock units are deposited and inserted and deformational processes change their shapes and locations.

Rock units are first emplaced either by deposition onto the surface or intrude into the overlying rock. Deposition can occur when sediments settle onto the surface of the Earth and later lithify into sedimentary rock, or when as volcanic material such as volcanic ash or lava flows, blanket the surface. Igneous intrusions such as batholiths, laccoliths, dikes, and sills, push upwards into the overlying rock, and crystallize as they intrude.

After the initial sequence of rocks has been deposited, the rock units can be deformed and/or metamorphosed. Deformation typically occurs as a result of horizontal shortening, horizontal extension, or side-to-side (strike-slip) motion. These structural regimes broadly relate to convergent boundaries, divergent boundaries, and transform boundaries, respectively, between tectonic plates.

## Historical perspective

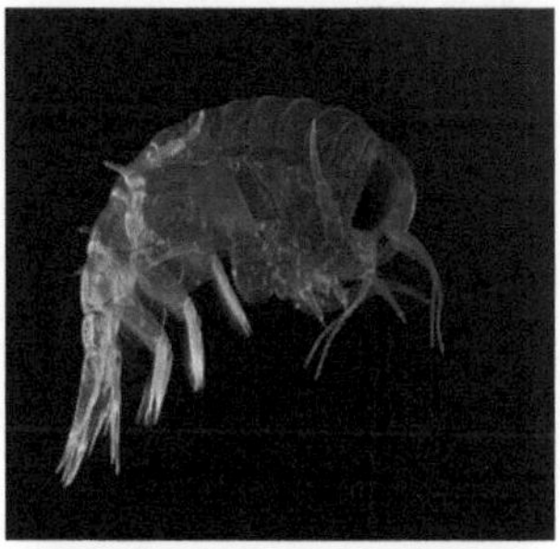

Plankton inhabit oceans, seas and lakes, and have existed in various forms for at least 2 billion years.[11]

Earth is estimated to have formed 4.54 billion years ago from the solar nebula, along with the Sun and other planets.[12] The moon formed roughly 20 million years later. Initially molten, the outer layer of the planet cooled, resulting in the solid crust. Outgassing and volcanic activity produced the primordial atmosphere. Condensing water vapor, most or all of which came from ice delivered by comets, produced the oceans and other water sources.[13] The highly energetic chemistry is believed to have produced a self-replicating molecule around 4 billion years ago.[14]

Continents formed, then broke up and reformed as the surface of Earth reshaped over hundreds of millions of years, occasionally combining to make a supercontinent. Roughly 750 million years ago, the earliest known supercontinent Rodinia, began to break apart. The continents later recombined to form Pannotia which broke apart about 540 million years ago, then finally Pangaea, which broke apart about 180 million years ago.[15]

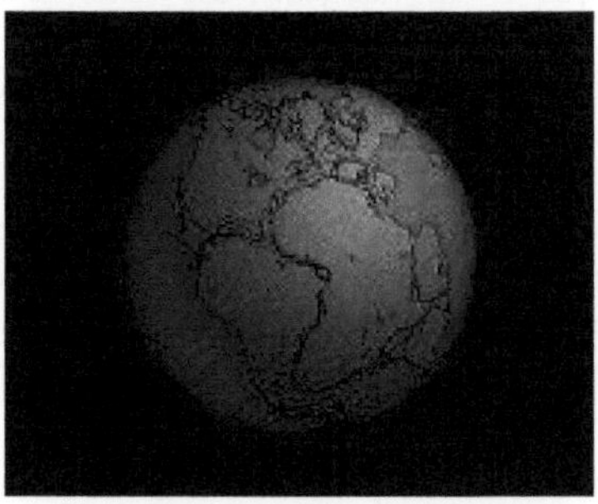

An animation of the Earth's hypothesized Pangaea separation.

There is significant evidence that a severe glacial action during the Neoproterozoic era covered much of the planet in a sheet of ice. This hypothesis has been termed the "Snowball Earth", and it is of particular interest as it precedes the Cambrian explosion in which multicellular life forms began to proliferate about 530–540 million years ago.[16]

Since the Cambrian explosion there have been five distinctly identifiable mass extinctions.[17] The last mass extinction occurred some 65 million years ago, when a meteorite collision probably triggered the extinction of the non-avian dinosaurs and other large reptiles, but spared small animals such as mammals, which then resembled shrews. Over the past 65 million years, mammalian life diversified.[18]

Several million years ago, a species of small African ape gained the ability to stand upright.[19] The subsequent advent of human life, and the development of agriculture and further civilization allowed humans to affect the Earth more rapidly than any previous life form, affecting both the nature and quantity of other organisms as well as global climate. By comparison, the Great Oxygenation Event, produced by the proliferation of algae during the Siderian period, required about 300 million years to culminate.

The present era is classified as part of a mass extinction event, the Holocene extinction event, the fastest ever to have occurred.[20] [21] Some, such as E. O. Wilson of Harvard University, predict that human destruction of the biosphere could cause the extinction of one-half of all species in the next 100 years.[22] The extent of the current extinction event is still being researched, debated and calculated by biologists.[23]

## Atmosphere, climate, and weather

Lightning

Blue light is scattered more than other wavelengths by the gases in the atmosphere, giving the Earth a blue halo when seen from space

The atmosphere of the Earth serves as a key factor in sustaining the planetary ecosystem. The thin layer of gases that envelops the Earth is held in place by the planet's gravity. Dry air consists of 78% nitrogen, 21% oxygen, 1% argon and other inert gases, carbon dioxide, etc.; but air also contains a variable amount of water vapor. The atmospheric pressure declines steadily with altitude, and has a scale height of about 8 kilometres at the Earth's surface: the height at which the atmospheric pressure has declined by a factor of $e$ (a mathematical constant equal to 2.71...).[24] [25] The ozone layer of the Earth's atmosphere plays an important role in depleting the amount of ultraviolet (UV) radiation that reaches the surface. As DNA is readily damaged by UV light, this serves to protect life at the surface. The atmosphere also retains heat during the night, thereby reducing the daily temperature extremes.

Terrestrial weather occurs almost exclusively in the lower part of the atmosphere, and serves as a convective system for redistributing heat. Ocean currents are another important factor in determining climate, particularly the major underwater thermohaline circulation which distributes heat energy from the equatorial oceans to the polar regions. These currents help to moderate the differences in temperature between winter and summer in the temperate zones. Also, without the redistributions of heat energy by the ocean currents and atmosphere, the tropics would be much hotter, and the polar regions much colder.

Weather can have both beneficial and harmful effects. Extremes in weather, such as tornadoes or hurricanes and cyclones, can expend large amounts of energy along their paths, and produce devastation. Surface vegetation has evolved a dependence on the seasonal variation of the weather, and sudden changes lasting only a few years can have a dramatic effect, both on the vegetation and on the animals which depend on its growth for their food.

A tornado in central Oklahoma

The planetary climate is a measure of the long-term trends in the weather. Various factors are known to influence the climate, including ocean currents, surface albedo, greenhouse gases, variations in the solar luminosity, and changes to the planet's orbit. Based on historical records, the Earth is known to have undergone drastic climate changes in the past, including ice ages.

The climate of a region depends on a number of factors, especially latitude. A latitudinal band of the surface with similar climatic attributes forms a climate region. There are a number of such regions, ranging from the tropical climate at the equator to the polar climate in the northern and southern extremes. Weather is also influenced by the seasons, which result from the Earth's axis being tilted relative to its orbital plane. Thus, at any given time during the summer or winter, one part of the planet is more directly exposed to the rays of the sun. This exposure alternates as the Earth revolves in its orbit. At any given time, regardless of season, the northern and southern hemispheres experience opposite seasons.

Weather is a chaotic system that is readily modified by small changes to the environment, so accurate weather forecasting is currently limited to only a few days. Overall, two things are currently happening worldwide: (1) temperature is increasing on the average; and (2) regional climates have been undergoing noticeable changes.[26]

## Water on Earth

The Iguazu Falls on the border between Brazil and Argentina

**Water** is a chemical substance that is composed of hydrogen and oxygen and is vital for all known forms of life.[27] In typical usage, *water* refers only to its liquid form or state, but the substance also has a solid state, ice, and a gaseous state, water vapor or steam. Water covers 71% of the Earth's surface.[28] On Earth, it is found mostly in oceans and other large water bodies, with 1.6% of water below ground in aquifers and 0.001% in the air as vapor, clouds (formed of solid and liquid water particles suspended in air), and precipitation.[29] Oceans hold 97% of surface water, glaciers and polar ice caps 2.4%, and other land surface water such as rivers, lakes and ponds 0.6%. Additionally, a minute amount of the Earth's water is contained within biological bodies and manufactured products.

## Oceans

A view of the Atlantic Ocean from Leblon, Rio de Janeiro.

An ocean is a major body of saline water, and a principal component of the hydrosphere. Approximately 71% of the Earth's surface (an area of some 361 million square kilometers) is covered by ocean, a continuous body of water that is customarily divided into several principal oceans and smaller seas. More than half of this area is over 3,000 meters (9,800 ft) deep. Average oceanic salinity is around 35 parts per thousand (ppt) (3.5%), and nearly all seawater has a salinity in the range of 30 to 38 ppt. Though generally recognized as several 'separate' oceans, these waters comprise one global, interconnected body of salt water often referred to as the World Ocean or global ocean.[30] [31] This concept of a global ocean as a continuous body of water with relatively free interchange among its parts is of fundamental importance to oceanography.[32]

The major oceanic divisions are defined in part by the continents, various archipelagos, and other criteria: these divisions are (in descending order of size) the Pacific Ocean, the Atlantic Ocean, the Indian Ocean, the Southern Ocean and the Arctic Ocean. Smaller regions of the oceans are called seas, gulfs, bays and other names. There are also salt lakes, which are smaller bodies of landlocked saltwater that are not interconnected with the World Ocean. Two notable examples of salt lakes are the Aral Sea and the Great Salt Lake.

## Lakes

Lake Mapourika, New Zealand

A lake (from Latin *lacus*) is a terrain feature (or physical feature), a body of liquid on the surface of a world that is localized to the bottom of basin (another type of landform or terrain feature; that is, it is not global) and moves slowly if it moves at all. On Earth, a body of water is considered a lake when it is inland, not part of the ocean, is larger and deeper than a pond, and is fed by a river.[33] [34] The only world other than Earth known to harbor lakes is Titan, Saturn's largest moon, which has lakes of ethane, most likely mixed with methane. It is not known if Titan's lakes are fed by rivers, though Titan's surface is carved by numerous river beds. Natural lakes on Earth are generally found in mountainous areas, rift zones, and areas with ongoing or recent glaciation. Other lakes are found in endorheic basins or along the courses of mature rivers. In some parts of the world, there are many lakes because of chaotic drainage patterns left over from the last Ice Age. All lakes are temporary over geologic time scales, as they will slowly fill in with sediments or spill out of the basin containing them.

### Ponds

A **pond** is a body of standing water, either natural or man-made, that is usually smaller than a lake. A wide variety of man-made bodies of water are classified as ponds, including water gardens designed for aesthetic ornamentation, fish ponds designed for commercial fish breeding, and solar ponds designed to store thermal energy. Ponds and lakes are distinguished from streams via current speed. While currents in streams are easily observed, ponds and lakes possess thermally driven microcurrents and moderate wind driven currents. These features distinguish a pond from many other aquatic terrain features, such as stream pools and tide pools.

The Westborough Reservoir (Mill Pond) in Westborough, Massachusetts.

## Rivers

The Nile river in Cairo, Egypt's capital city

A river is a natural watercourse,[35] usually freshwater, flowing toward an ocean, a lake, a sea or another river. In a few cases, a river simply flows into the ground or dries up completely before reaching another body of water. Small rivers may also be called by several other names, including stream, creek, brook, rivulet, and rill; there is no general rule that defines what can be called a river. Many names for small rivers are specific to geographic location; one example is *Burn* in Scotland and North-east England. Sometimes a river is said to be larger than a creek,[36] but this is not always the case, due to vagueness in the language.[37] A river is part of the hydrological cycle. Water within a river is generally collected from precipitation through surface runoff, groundwater recharge, springs, and the release of stored water in natural ice and snowpacks (i.e., from glaciers).

## Streams

A stream is a flowing body of water with a current, confined within a bed and stream banks. In the United States a stream is classified as a watercourse less than 60 feet (18 metres) wide. Streams are important as conduits in the water cycle, instruments in groundwater recharge, and they serve as corridors for fish and wildlife migration. The biological habitat in the immediate vicinity of a stream is called a riparian zone. Given the status of the ongoing Holocene extinction, streams play an important corridor role in connecting fragmented habitats and thus in conserving biodiversity. The study of streams and waterways in general involves many branches of inter-disciplinary natural science and engineering, including hydrology, fluvial geomorphology, aquatic ecology, fish biology, riparian ecology and others.

A rocky stream in Hawaii

## Ecosystems

Loch Lomond in Scotland forms a relatively isolated ecosystem. The fish community of this lake has remained unchanged over a very long period of time.[38]

An aerial view of a human ecosystem. Pictured is the city of Chicago

Ecosystems are composed of a variety of abiotic and biotic components that function in an interrelated way.[39] The structure and composition is determined by various environmental factors that are interrelated. Variations of these factors will initiate dynamic modifications to the ecosystem. Some of the more important components are: soil, atmosphere, radiation from the sun, water, and living organisms.

Central to the ecosystem concept is the idea that living organisms interact with every other element in their local environment. Eugene Odum, a founder of ecology, stated: "Any unit that includes all of the organisms (ie: the "community") in a given area interacting with the physical environment so that a flow of energy leads to clearly defined trophic structure, biotic diversity, and material cycles (i.e.: exchange of materials between living and nonliving parts) within the system is an ecosystem."[40] Within the ecosystem, species are connected and dependent upon one another in the food chain, and exchange energy and matter between themselves as well as with their environment.[41] The human ecosystem concept is grounded in the deconstruction of the human/nature dichotomy and the premise that all species are ecologically integrated with each other, as well as with the abiotic constituents of their biotope.

A smaller unit of size is called a microecosystem. For example, a microsystem can be a stone and all the life under it. A *macroecosystem* might involve a whole ecoregion, with its drainage basin.[42]

## Wilderness

Old growth European Beech forest in Biogradska Gora National Park, Montenegro.

**Wilderness** is generally defined as areas that have not been significantly modified by human activity. The WILD Foundation [43] goes into more detail, defining wilderness as: "The most intact, undisturbed wild natural areas left on our planet – those last truly wild places that humans do not control and have not developed with roads, pipelines or other industrial infrastructure." Wilderness areas can be found in preserves, estates, farms, conservation preserves, ranches, national forests, national parks and even in urban areas along rivers, gulches or otherwise undeveloped areas. Wilderness areas [43] and protected parks are considered important for the survival of certain species, ecological studies, conservation, solitude, and recreation. Some nature writers believe wilderness areas are vital for the human spirit and creativity,[44] and some Ecologists consider wilderness areas to be an integral part of the planet's self-sustaining natural ecosystem (the biosphere). They may also preserve historic genetic traits and that they provide habitat for wild flora and fauna that may be difficult to recreate in zoos, arboretums or laboratories.

## Life

Female mallard and ducklings – reproduction is essential for continuing life

Although there is no universal agreement on the definition of life, scientists generally accept that the biological manifestation of life is characterized by organization, metabolism, growth, adaptation, response to stimuli and reproduction.[45] Life may also be said to be simply the characteristic state of organisms.

Properties common to terrestrial organisms (plants, animals, fungi, protists, archaea and bacteria) are that they are cellular, carbon-and-water-based with complex organization, having a metabolism, a capacity to grow, respond to stimuli, and reproduce. An entity with these properties is generally considered life. However, not every definition of life considers all of these properties to be essential. Human-made analogs of life may also be considered to be life.

The biosphere is the part of Earth's outer shell – including land, surface rocks, water, air and the atmosphere – within which life occurs, and which biotic processes in turn alter or transform. From the broadest geophysiological point of view, the biosphere is the global ecological system integrating all living beings and their relationships, including their interaction with the elements of the lithosphere (rocks), hydrosphere (water), and atmosphere (air). Currently the entire Earth contains over 75 billion tons (150 *trillion* pounds or about $6.8 \times 10^{13}$ kilograms) of biomass (life), which lives within various environments within the biosphere.[46]

Over nine-tenths of the total biomass on Earth is plant life, on which animal life depends very heavily for its existence.[47] More than 2 million species of plant and animal life have been identified to date,[48] and estimates of the actual number of existing species range from several million to well over 50 million.[49] [50] [51] The number of individual species of life is constantly in some degree of flux, with new species appearing and others ceasing to exist on a continual basis.[52] [53] The total number of species is presently in rapid decline.[54] [55] [56]

### Evolution

Life is only known to exist on the planet Earth.(cf Astrobiology) The origin of life is still a poorly understood process, but it is thought to have occurred about 3.9 to 3.5 billion years ago during the hadean or archean eons on a primordial earth that had a substantially different environment than is found at present.[59] These life forms possessed the basic traits of self-replication and inheritable traits. Once life had appeared, the process of evolution by natural selection resulted in the development of ever-more diverse life forms.

An area of the Amazon Rainforest in Brazil. The tropical rainforests of South America contain the largest diversity of species on Earth.[57] [58]

Species that were unable to adapt to the changing environment and competition from other life forms became extinct. However, the fossil record retains evidence of many of these older species. Current fossil and DNA evidence shows that all existing species can trace a continual ancestry back to the first primitive life forms.[59]

The advent of photosynthesis in very basic forms of plant life worldwide allowed the sun's energy to be harvested to create conditions allowing for more complex life. The resultant oxygen accumulated in the atmosphere and gave rise to the ozone layer. The incorporation of smaller cells within larger ones resulted in the development of yet more complex cells called eukaryotes.[60] Cells within colonies became increasingly specialized, resulting in true multicellular organisms. With the ozone layer absorbing harmful ultraviolet radiation, life colonized the surface of

Earth.

## Microbes

The first form of life to develop on the Earth were microbes, and they remained the only form of life on the planet until about a billion years ago when multi-cellular organisms began to appear.[61] Microorganisms are single-celled organisms that are generally microscopic, and smaller than the human eye can see. They include Bacteria, Fungi, Archaea and Protista.

A microscopic mite *Lorryia formosa.*

These life forms are found in almost every location on the Earth where there is liquid water, including the interior of rocks within the planet.[62] Their reproduction is both rapid and profuse. The combination of a high mutation rate and a horizontal gene transfer[63] ability makes them highly adaptable, and able to survive in new environments, including outer space.[64] They form an essential part of the planetary ecosystem. However some microorganisms are pathogenic and can post health risk to other organisms.

## Plants and animals

A selection of diverse plant species

Originally Aristotle divided all living things between plants, which generally do not move fast enough for humans to notice, and animals. In Linnaeus' system, these became the kingdoms Vegetabilia (later Plantae) and Animalia. Since then, it has become clear that the Plantae as originally defined included several unrelated groups, and the fungi and several groups of algae were removed to new kingdoms. However, these are still often considered plants in many contexts. Bacterial life is sometimes included in flora,[65] [66] and some classifications use the term *bacterial flora* separately from *plant flora.*

Among the many ways of classifying plants are by regional floras, which, depending on the purpose of study, can also include *fossil flora,* remnants

of plant life from a previous era. People in many regions and countries take great pride in their individual arrays of characteristic flora, which can vary widely across the globe due to differences in climate and terrain.

Regional floras commonly are divided into categories such as *native flora* and *agricultural and garden flora,* the lastly mentioned of which are intentionally grown and cultivated. Some types of "native flora" actually have been introduced centuries ago by people migrating from one region or continent to another, and become an integral part of the native, or natural flora of the place to which they were introduced. This is an example of how human interaction with nature can blur the boundary of what is considered nature.

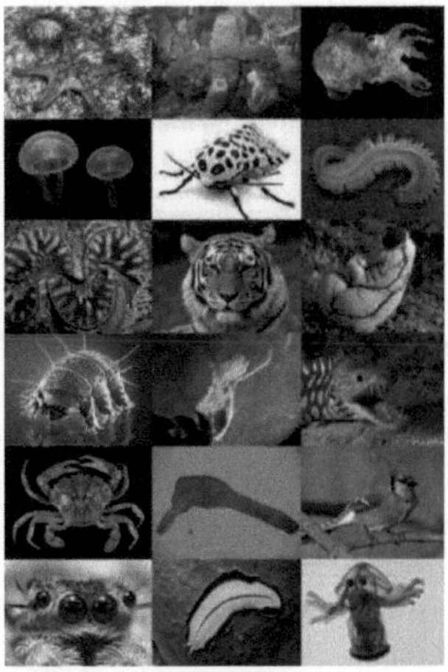

There are many animal species on the planet

Another category of plant has historically been carved out for *weeds*. Though the term has fallen into disfavor among botanists as a formal way to categorize "useless" plants, the informal use of the word "weeds" to describe those plants that are deemed worthy of elimination is illustrative of the general tendency of people and societies to seek to alter or shape the course of nature. Similarly, animals are often categorized in ways such as *domestic*, *farm animals*, *wild animals*, *pests*, etc. according to their relationship to human life.

Animals as a category have several characteristics that generally set them apart from other living things, though this is not traced by scientists to having legs or wings instead of roots and leaves. Animals are eukaryotic and usually multicellular (although see Myxozoa), which separates them from bacteria, archaea and most protists. They are heterotrophic, generally digesting food in an internal chamber, which separates them from plants and algae. They are also distinguished from plants, algae, and fungi by lacking cell walls.

With a few exceptions, most notably the sponges (Phylum Porifera), animals have bodies differentiated into separate tissues. These include muscles, which are able to contract and control locomotion, and a nervous system, which sends and processes signals. There is also typically an internal digestive chamber. The eukaryotic cells possessed by all animals are surrounded by a characteristic extracellular matrix composed of collagen and elastic glycoproteins. This may be calcified to form structures like shells, bones, and spicules, a framework upon which cells can move about and be reorganized during development and maturation, and which supports the complex anatomy required for mobility.

## Human interrelationship

Despite their natural beauty, the secluded valleys along the Na Pali Coast in Hawaii are heavily modified by introduced invasive species such as She-oak.

Although humans currently comprise only a minuscule proportion of the total living biomass on Earth, the human effect on nature is disproportionately large. Because of the extent of human influence, the boundaries between what humans regard as nature and "made environments" is not clear cut except at the extremes. Even at the extremes, the amount of natural environment that is free of discernible human influence is presently diminishing at an increasingly rapid pace.

The development of technology by the human race has allowed the greater exploitation of natural resources and has helped to alleviate some of the risk from natural hazards. In spite of this progress, however, the fate of human civilization remains closely linked to changes in the environment. There exists a highly complex feedback loop between the use of advanced technology and changes to the environment that are only slowly becoming understood.[67] Man-made threats to the Earth's natural environment include pollution, deforestation, and disasters such as oil spills. Humans have contributed to the extinction of many plants and animals.

Sochi dendrarium is an example of confluence of "natural" and a "made" environment

Humans employ nature for both leisure and economic activities. The acquisition of natural resources for industrial use remains the primary component of the world's economic system. Some activities, such as hunting and fishing, are used for both sustenance and leisure, often by different people. Agriculture was first adopted around the 9th millennium BCE. Ranging from food production to energy, nature influences economic wealth.

Although early humans gathered uncultivated plant materials for food and employed the medicinal properties of vegetation for healing,[68] most modern human use of plants is through agriculture. The clearance of large tracts of land for crop growth has led to a significant reduction in the amount available of forestation and wetlands, resulting in the loss of habitat for many plant and animal species as well as increased erosion.[69]

## Aesthetics and beauty

*Pinguicula grandiflora*, commonly known as a Butterwort

Beauty in nature has historically been a prevalent theme in art and books, filling large sections of libraries and bookstores. That nature has been depicted and celebrated by so much art, photography, poetry and other literature shows the strength with which many people associate nature and beauty. Reasons why this association exists, and what the association consists of, is studied by the branch of philosophy called aesthetics. Beyond certain basic characteristics that many philosophers agree about to explain what is seen as beautiful, the opinions are virtually endless.[70] Nature and wildness have been important subjects in various eras of world history. An early tradition of landscape art began in China during the Tang Dynasty (618–907). The tradition of representing nature *as it is* became one of the aims of Chinese painting and was a significant influence in Asian art.

Although natural wonders are celebrated in the Psalms and the Book of Job, wilderness portrayals in art became more prevalent in the 1800s, especially in the works of the Romantic movement. British artists John Constable and J. M. W. Turner turned their attention to capturing the beauty of the natural world in their paintings. Before that, paintings had been primarily of religious scenes or of human beings. William Wordsworth's poetry described the wonder of the natural world, which had formerly been viewed as a threatening place. Increasingly the valuing of nature became an aspect of Western culture.[71] This artistic movement also coincided with the Transcendentalist movement in the Western world. A common classical idea of beautiful art involves the word mimesis, the imitation of nature. Also in the realm of ideas about beauty in nature is that the perfect is implied through symmetry, equal division, and other perfect mathematical forms and notions.

## Matter and energy

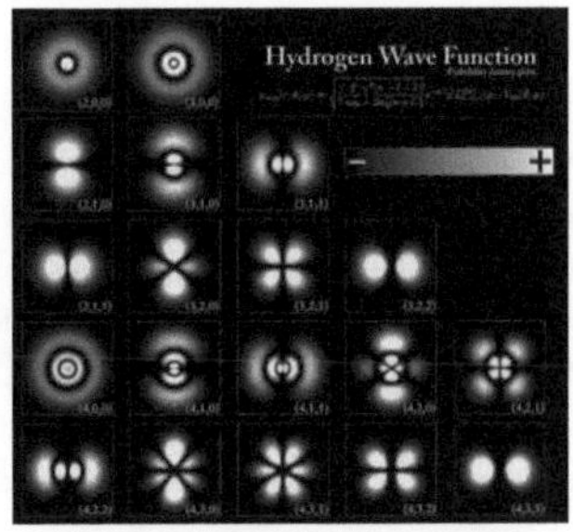

The first few hydrogen atom electron orbitals shown as cross-sections with color-coded probability density

Some fields of science see nature as matter in motion, obeying certain laws of nature which science seeks to understand. For this reason the most fundamental science is generally understood to be "physics" – the name for which is still recognizable as meaning that it is the study of nature.

Matter is commonly defined as the substance of which physical objects are composed. It constitutes the observable universe. The visible components of the universe are now believed to compose only 4 percent of the total mass. The remainder is believed to consist of 23 percent cold dark matter and 73 percent dark energy.[72] The exact nature of these components is still unknown and is currently under intensive investigation by physicists.

The behavior of matter and energy throughout the observable universe appears to follow well-defined physical laws. These laws have been employed to produce cosmological models that successfully explain the structure and the evolution of the universe we can observe. The mathematical expressions of the laws of physics employ a set of twenty physical constants[73] that appear to be static across the observable universe.[74] The values of these constants have been carefully measured, but the reason for their specific values remains a mystery.

## Beyond Earth

Planets and dwarf planets of the Solar System *(Sizes to scale, distances not to scale)*

Outer space, also simply called *space*, refers to the relatively empty regions of the universe outside the atmospheres of celestial bodies. *Outer* space is used to distinguish it from airspace (and terrestrial locations). There is no discrete boundary between the Earth's atmosphere and space, as the atmosphere gradually attenuates with increasing altitude. Outer space within the Solar System is called interplanetary space, which passes over into interstellar space at what is known as the heliopause.

Outer space is certainly spacious, but it is far from empty. Outer space is sparsely filled with several dozen types of organic molecules discovered to date by microwave spectroscopy, blackbody radiation left over from the big bang and the origin of the universe, and cosmic rays, which include ionized atomic nuclei and various subatomic particles. There is also some gas, plasma and dust, and small meteors. Additionally, there are signs of human life in outer space today, such as material left over from previous manned and unmanned launches which are a potential hazard to spacecraft. Some of this debris re-enters the atmosphere periodically.

**NGC 4414** is a spiral galaxy in the constellation Coma Berenices about 56,000 light years in diameter and approximately 60 million light years from Earth

Although the planet Earth is currently the only known body within the solar system to support life, current evidence suggests that in the distant past the planet Mars possessed bodies of liquid water on the surface.[75] For a brief period in Mars' history, it may have also been capable of forming life. At present though, most of the water remaining on Mars is frozen. If life exists at all on Mars, it is most likely to be located underground where liquid water can still exist.[76]

Conditions on the other terrestrial planets, Mercury and Venus, appear to be too harsh to support life as we know it. But it has been conjectured that Europa, the fourth-largest moon of Jupiter, may possess a sub-surface ocean of liquid water and could potentially host life.[77]

Recently, the team of Stéphane Udry have discovered a new planet named Gliese 581 g, which is an extrasolar planet orbiting the red dwarf star Gliese 581. Gliese 581 g appears to lie in the habitable zone of space surrounding the star, and therefore could possibly host life as we know it.

## See also

**Media:**

- *Natural History*, by Pliny the Elder
- *Nature*, by Ralph Waldo Emerson
- *Nature*, a prominent scientific journal
- *National Wildlife (magazine)*, publication of the National Wildlife Federation
- Nature (TV series)

**Organizations:**

- The Nature Conservancy
- Nature Detectives

**Science:**

- Natural history
- Natural landscape

**Philosophy:**

- Mother Nature
- Nature (philosophy)
- Naturalism (philosophy): any of several philosophical stances, typically those descended from Materialism and Pragmatism that do not distinguish the supernatural from nature. This includes the *methodological naturalism* of natural science, which makes the methodological assumption that observable events in nature are explained only by natural causes, without assuming either the existence or non-existence of the supernatural.
- Balance of nature (biological fallacy): A discredited concept of natural equilibrium in predator:prey dynamics.

## Notes and references


[1] Harper, Douglas. "nature" (http://www.etymonline.com/index.php?term=nature). *Online Etymology Dictionary*. . Retrieved 2006-09-23.

[2] A useful though somewhat erratically presented account of the pre-Socratic use of the concept of φύσις may be found in Naddaf, Gerard *The Greek Concept of Nature*, SUNY Press, 2006. The word φύσις, while first used in connection with a plant in Homer, occurs very early in Greek philosophy, and in several senses. Generally, these senses match rather well the current senses in which the English word *nature* is used, as confirmed by Guthrie, W.K.C. *Presocratic Tradition from Parmenides to Democritus* (volume 2 of his *History of Greek Philosophy*), Cambridge UP, 1965.

[3] The first known use of *physis* was by Homer in reference to the intrinsic qualities of a plant: ὣς ἄρα φωνήσας πόρε φάρμακον ἀργεϊφόντης ἐκ γαίης ἐρύσας, καί μοι **φύσιν** αὐτοῦ ἔδειξε. (So saying, Argeiphontes [=Hermes] gave me the herb, drawing it from the ground, and showed me its **nature**.) *Odyssey* 10.302-3 (ed. A.T. Murray). (The word is dealt with thoroughly in Liddell and Scott's *Greek Lexicon* (http://archimedes.fas.harvard.edu/pollux).) For later but still very early Greek uses of the term, see earlier note.

[4] Isaac Newton's Philosophiae Naturalis Principia Mathematica (1687), for example, is translated "Mathematical Principles of Natural Philosophy", and reflects the then-current use of the words "natural philosophy", akin to "systematic study of nature"

[5] The etymology of the word "physical" shows its use as a synonym for "natural" in about the mid-15th century: Harper, Douglas. "physical" (http://www.etymonline.com/index.php?term=physical). *Online Etymology Dictionary*. . Retrieved 2006-09-20.

[6] "World Climates" (http://www.blueplanetbiomes.org/climate.htm). *Blue Planet Biomes*. . Retrieved 2006-09-21.

[7] "Calculations favor reducing atmosphere for early Earth" (http://www.sciencedaily.com/releases/2005/09/050911103921.htm). Science Daily. 2005-09-11. . Retrieved 2007-01-06.

[8] "Past Climate Change" (http://www.epa.gov/climatechange/science/pastcc.html). U.S. Environmental Protection Agency. . Retrieved 2007-01-07.

[9] Hugh Anderson, Bernard Walter (March 28, 1997). "History of Climate Change" (http://web.archive.org/web/20080123130745/http://vathena.arc.nasa.gov/curric/land/global/climchng.html). NASA. Archived from the original (http://vathena.arc.nasa.gov/curric/land/global/climchng.html) on 2008-01-23. . Retrieved 2007-01-07.

[10] Weart, Spencer (June 2006). "The Discovery of Global Warming" (http://www.aip.org/history/climate/). American Institute of Physics. . Retrieved 2007-01-07.

[11] Margulis, Lynn; Dorian Sagan (1995). *What is Life?*. New York: Simon & Schuster. ISBN 0684813262.

[12] Dalrymple, G. Brent (1991). *The Age of the Earth*. Stanford: Stanford University Press. ISBN 0-8047-1569-6.

[13] Morbidelli, A.; *et al.* (2000). "Source Regions and Time Scales for the Delivery of Water to Earth". *Meteoritics & Planetary Science* **35** (6): 1309–1320. Bibcode 2000M&PS...35.1309M. doi:10.1111/j.1945-5100.2000.tb01518.x.

[14] "Earth's Oldest Mineral Grains Suggest an Early Start for Life" (http://nai.arc.nasa.gov/news_stories/news_detail.cfm?ID=76). NASA Astrobilogy Institute. 2001-12-24. . Retrieved 2006-05-24.

[15] Murphy, J.B.; R.D. Nance (2004). "How do supercontinents assemble?" (http://www.americanscientist.org/issues/page2/how-do-supercontinents-assemble). *American Scientist* **92** (4): 324. doi:10.1511/2004.4.324. .

[16] Kirschvink, J.L. (1992). "Late Proterozoic Low-Latitude Global Glaciation: The Snowball Earth" (http://www.gps.caltech.edu/~jkirschvink/pdfs/firstsnowball.pdf). In J.W. Schopf, C. Klein eds.. *The Proterozoic Biosphere*. Cambridge: Cambridge University Press. pp. 51–52. ISBN 0-521-36615-1. .

[17] Raup, David M.; J. John Sepkoski Jr. (March 1982). "Mass extinctions in the marine fossil record". *Science* **215** (4539): 1501–3. Bibcode 1982Sci...215.1501R. doi:10.1126/science.215.4539.1501. PMID 17788674.

[18] Margulis, Lynn; Dorian Sagan (1995). *What is Life?*. New York: Simon & Schuster. p. 145. ISBN 0-684-81326-2.

[19] Margulis, Lynn; Dorian Sagan (1995). *What is Life?*. New York: Simon & Schuster. ISBN 0-684-81326-2.

[20] Diamond J; Ashmole, N. P.; Purves, P. E. (1989). "The present, past and future of human-caused extinctions". *Philos Trans R Soc Lond B Biol Sci* **325** (1228): 469–76; discussion 476–7. Bibcode 1989RSPTB.325..469D. doi:10.1098/rstb.1989.0100. PMID 2574887.

[21] Novacek M, Cleland E (2001). "The current biodiversity extinction event: scenarios for mitigation and recovery". *Proc Natl Acad Sci USA* **98** (10): 5466–70. Bibcode 2001PNAS...98.5466N. doi:10.1073/pnas.091093698. PMC 33235. PMID 11344295.

[22] Wick, Lucia; Möhl, Adrian (2006). "The mid-Holocene extinction of silver fir (Abies alba) in the Southern Alps: a consequence of forest fires? Palaeobotanical records and forest simulations". *Vegetation History and Archaeobotany* **15** (4): 435–444. doi:10.1007/s00334-006-0051-0.

[23] See, e.g. (http://park.org/Canada/Museum/extinction/holmass.html), (http://park.org/Canada/Museum/extinction/extincmenu.html), (http://park.org/Canada/Museum/extinction/patterns.html)

[24] "Ideal Gases under Constant Volume, Constant Pressure, Constant Temperature, & Adiabatic Conditions" (http://www.grc.nasa.gov/WWW/K-12/Numbers/Math/Mathematical_Thinking/ideal_gases_under_constant.htm). NASA. . Retrieved 2007-01-07.

[25] Pelletier, Jon D. (2002). "Natural variability of atmospheric temperatures and geomagnetic intensity over a wide range of time scales". *Proceedings of the National Academy of Sciences* **99** (90001): 2546–2553. Bibcode 2002PNAS...99.2546P. doi:10.1073/pnas.022582599. PMC 128574. PMID 11875208.

[26] "Tropical Ocean Warming Drives Recent Northern Hemisphere Climate Change" (http://www.sciencedaily.com/releases/2001/04/010406073554.htm). Science Daily. April 6, 2001. . Retrieved 2006-05-24.

[27] "Water for Life" (http://www.un.org/waterforlifedecade/background.html). Un.org. 2005-03-22. . Retrieved 2011-05-14.

[28] "CIA- The world fact book" (https://www.cia.gov/library/publications/the-world-factbook/geos/xx.html#Geo). Central Intelligence Agency. . Retrieved 2008-12-20.

[29] Water Vapor in the Climate System (http://www.agu.org/sci_soc/mockler.html), Special Report, [AGU], December 1995 (linked 4/2007). Vital Water (http://www.unep.org/dewa/assessments/ecosystems/water/vitalwater/) UNEP.

[30] " Ocean (http://www.answers.com/Ocean#Encyclopedia)". *The Columbia Encyclopedia.* 2002. New York: Columbia University Press

[31] " Distribution of land and water on the planet (http://www.oceansatlas.com/unatlas/about/physicalandchemicalproperties/background/seemore1.html)". *UN Atlas of the Oceans* (http://www.oceansatlas.com/)

[32] Spilhaus, Athelstan F. 1942 (Jul.). "Maps of the whole world ocean." *Geographical Review* (American Geographical Society). Vol. 32 (3): pp. 431–5.

[33] Brittanica online. "Lake (physical feature)" (http://www.britannica.com/EBchecked/topic/328083/lake). . Retrieved 2008-06-25. "[a Lake is] any relatively large body of slowly moving or standing water that occupies an inland basin of appreciable size. Definitions that precisely distinguish lakes, ponds, swamps, and even rivers and other bodies of nonoceanic water are not well established. It may be said, however, that rivers and streams are relatively fast moving; marshes and swamps contain relatively large quantities of grasses, trees, or shrubs; and ponds are relatively small in comparison to lakes. Geologically defined, lakes are temporary bodies of water."

[34] a body of fresh or salt water of considerable size, surrounded by land. "Dictionary.com definition" (http://dictionary.reference.com/browse/lake). a body of fresh or salt water of considerable size, surrounded by land.. Retrieved 2008-06-25.

[35] River {definition} (http://www.merriam-webster.com/dictionary/river) from Merriam-Webster. Accessed February 2010.

[36] River (http://wordnet.princeton.edu/perl/webwn?s=river&sub=Search+WordNet&o2=&o0=1&o7=&o5=&o1=1&o6=&o4=&o3=&h=), Wordnet

[37] USGS – U.S. Geological Survey – faqs (http://geonames.usgs.gov/domestic/faqs.htm), #17 What is the difference between mountain, hill, and peak; lake and pond; or river and creek?

[38] Adams, C.E. (1994). "The fish community of Loch Lomond, Scotland : its history and rapidly changing status" (http://cat.inist.fr/?aModele=afficheN&cpsidt=3302548). *Hydrobiologia* **290** (1–3): 91–102. doi:10.1007/BF00008956. .

[39] Pidwirny, Michael (2006). "Introduction to the Biosphere: Introduction to the Ecosystem Concept" (http://www.physicalgeography.net/fundamentals/9j.html). *Fundamentals of Physical Geography (2nd Edition).* . Retrieved September 28, 2006.

[40] Odum, EP (1971) Fundamentals of ecology, third edition, Saunders New York

[41] Pidwirny, Michael (2006). "Introduction to the Biosphere: Organization of Life" (http://www.physicalgeography.net/fundamentals/9d.html). *Fundamentals of Physical Geography (2nd Edition).* . Retrieved September 28, 2006.

[42] Bailey, Robert G. (April 2004). "Identifying Ecoregion Boundaries" (http://www.fs.fed.us/institute/news_info/Identifying_ecoregion_boundaries.pdf) (PDF). *Environmental Management* **34** (Supplement 1): S14–26. doi:10.1007/s00267-003-0163-6. PMID 15883869. .

[43] http://www.wild.org/main/about/what-is-a-wilderness-area/

[44] No Man's Garden by Daniel B. Botkin p155-157

[45] "Definition of Life" (http://www.calacademy.org/exhibits/xtremelife/what_is_life.php). California Academy of Sciences. 2006. . Retrieved 2007-01-07.

[46] The figure "about one-half of one percent" takes into account the following (See, e.g., Leckie, Stephen (1999). "How Meat-centred Eating Patterns Affect Food Security and the Environment" (http://www.idrc.ca/en/ev-30610-201-1-DO_TOPIC.html). *For hunger-proof cities : sustainable urban food systems.* Ottawa: International Development Research Centre. ISBN 0-88936-882-1. ., which takes global average weight as 60 kg.), the total human biomass is the average weight multiplied by the current human population of approximately 6.5 billion (see, *e.g.*, "World Population Information" (http://www.census.gov/ipc/www/world.html). U.S. Census Bureau. . Retrieved September 28, 2006.): Assuming 60–70 kg to be the average human mass (approximately 130–150 lb on the average), an approximation of total global human mass of between 390 billion ($390\times10^9$) and 455 billion kg (between 845 billion and 975 billion lb, or about 423 million–488 million short tons). The total biomass of all kinds on earth is estimated to be in excess of $6.8 \times 10^{13}$ kg (75 billion short tons). By these calculations, the portion of total biomass accounted for by humans would be very roughly 0.6%.

[47] Sengbusch, Peter V.. "The Flow of Energy in Ecosystems – Productivity, Food Chain, and Trophic Level" (http://www.biologie.uni-hamburg.de/b-online/e54/54c.htm). *Botany online.* University of Hamburg Department of Biology. . Retrieved September 23, 2006.

[48] Pidwirny, Michael (2006). "Introduction to the Biosphere: Species Diversity and Biodiversity" (http://www.physicalgeography.net/fundamentals/9h.html). *Fundamentals of Physical Geography (2nd Edition).* . Retrieved September 23, 2006.

[49] "How Many Species are There?" (http://faculty.plattsburgh.edu/thomas.wolosz/howmanysp.htm). *Extinction Web Page Class Notes.* . Retrieved September 23, 2006.

[50] "Animal." World Book Encyclopedia. 16 vols. Chicago: World Book, 2003. This source gives an estimate of from 2 to 50 million.

[51] "Just How Many Species Are There, Anyway?" (http://www.sciencedaily.com/releases/2003/05/030526103731.htm). Science Daily. May 2003. . Retrieved September 26, 2006.

[52] Withers, Mark A.; et al. (1998). "Changing Patterns in the Number of Species in North American Floras" (http://biology.usgs.gov/luhna/chap4.html). *Land Use History of North America.* . Retrieved September 26, 2006. Website based on the contents of the book: Sisk, T.D., ed., ed (1998). *Perspectives on the land use history of North America: a context for understanding our changing environment* (Revised September 1999 ed.). U.S. Geological Survey, Biological Resources Division. USGS/BRD/BSR-1998-0003.

[53] "Tropical Scientists Find Fewer Species Than Expected" (http://www.sciencedaily.com/releases/2002/04/020425072847.htm). Science Daily. April 2002. . Retrieved September 27, 2006.

[54] Bunker, Daniel E.; et al. (November 2005). "Species Loss and Aboveground Carbon Storage in a Tropical Forest" (http://www.sciencemag.org/cgi/content/abstract/310/5750/1029). *Science* **310** (5750): 1029–31. Bibcode 2005Sci...310.1029B. doi:10.1126/science.1117682. PMID 16239439. .

[55] Wilcox, Bruce A. (2006). "Amphibian Decline: More Support for Biocomplexity as a Research Paradigm". *EcoHealth* **3** (1): 1. doi:10.1007/s10393-005-0013-5.

[56] Clarke, Robin, Robert Lamb, Dilys Roe Ward eds., ed (2002). "Decline and loss of species" (http://www.grida.no/geo/geo3/english/221.htm). *Global environment outlook 3 : past, present and future perspectives.* London; Sterling, VA: Nairobi, Kenya : UNEP. ISBN 92-807-2087-2. .

[57] "Why the Amazon Rainforest is So Rich in Species : News" (http://earthobservatory.nasa.gov/Newsroom/view.php?id=28907). Earthobservatory.nasa.gov. 2005-12-05. . Retrieved 2011-05-14.

[58] "Why The Amazon Rainforest Is So Rich In Species" (http://www.sciencedaily.com/releases/2005/12/051205163236.htm). Sciencedaily.com. 2005-12-05. . Retrieved 2011-05-14.

[59] Line M (1 January 2002). "The enigma of the origin of life and its timing" (http://mic.sgmjournals.org/cgi/content/full/148/1/21?view=long&pmid=11782495). *Microbiology* **148** (Pt 1): 21–7. PMID 11782495. .

[60] Berkner, L. V.; L. C. Marshall (May 1965). "On the Origin and Rise of Oxygen Concentration in the Earth's Atmosphere". *Journal of the Atmospheric Sciences* **22** (3): 225–261. Bibcode 1965JAtS...22..225B. doi:10.1175/1520-0469(1965)022<0225:OTOARO>2.0.CO;2. ISSN 1520-0469.

[61] Schopf J (1994). "Disparate rates, differing fates: tempo and mode of evolution changed from the Precambrian to the Phanerozoic". *Proc Natl Acad Sci USA* **91** (15): 6735–42. Bibcode 1994PNAS...91.6735S. doi:10.1073/pnas.91.15.6735. PMC 44277. PMID 8041691.

[62] Szewzyk U, Szewzyk R, Stenström T (1994). "Thermophilic, anaerobic bacteria isolated from a deep borehole in granite in Sweden". *Proc Natl Acad Sci USA* **91** (5): 1810–3. Bibcode 1994PNAS...91.1810S. doi:10.1073/pnas.91.5.1810. PMC 43253. PMID 11607462.

[63] Wolska K (2003). "Horizontal DNA transfer between bacteria in the environment". *Acta Microbiol Pol* **52** (3): 233–43. PMID 14743976.

[64] Horneck G (1981). "Survival of microorganisms in space: a review". *Adv Space Res* **1** (14): 39–48. doi:10.1016/0273-1177(81)90241-6. PMID 11541716.

[65] "flora" (http://webster.com/cgi-bin/dictionary?va=flora). *Merriam-Webster Online Dictionary.* Merriam-Webster. . Retrieved September 27, 2006.

[66] "Glossary" (http://biology.usgs.gov/s+t/SNT/noframe/zy198.htm#F). *Status and Trends of the Nation's Biological Resources.* Reston, VA: Department of the Interior, Geological Survey. 1998. SuDocs No. I 19.202:ST 1/V.1-2. .

[67] "Feedback Loops In Global Climate Change Point To A Very Hot 21st Century" (http://www.lbl.gov/Science-Articles/Archive/ESD-feedback-loops.html). Science Daily. May 22, 2006. . Retrieved 2007-01-07.

[68] "Plant Conservation Alliance – Medicinal Plant Working Groups Green Medicine" (http://www.nps.gov/plants/medicinal/plants.htm). US National Park Services. . Retrieved September 23, 2006.

[69] Oosthoek, Jan (1999). "Environmental History: Between Science & Philosophy" (http://www.eh-resources.org/philosophy.html). Environmental History Resources. . Retrieved 2006-12-01.

[70] For an example of a range of opinions, see: "On the Beauty of Nature" (http://www.wilderness.org/Library/Documents/Beauty_Quotes.cfm). The Wilderness Society. . Retrieved September 29, 2006. and Ralph Waldo Emerson's analysis of the subject: Emerson, Ralph Waldo (1849). "Beauty" (http://www.emersoncentral.com/beauty.htm). *Nature; Addresses and Lectures.* .

[71] History of Conservation (http://www.spacesfornature.org/greatspaces/conservation.html) BC Spaces for Nature. Accessed: May 20, 2006.

[72] "Some Theories Win, Some Lose" (http://map.gsfc.nasa.gov/m_mm/mr_limits.html). *WMAP Mission: First Year Results.* NASA. . Retrieved 29 2006.

[73] Taylor, Barry N. (1971). "Introduction to the constants for nonexperts" (http://www.physics.nist.gov/cuu/Constants/introduction.html). National Institute of Standards and Technology. . Retrieved 2007-01-07.

[74] D. A. Varshalovich, A. Y. Potekhin, A. V. Ivanchik (2000). "Testing cosmological variability of fundamental constants". *AIP Conference Proceedings* **506**: 503. arXiv:physics/0004062. doi:10.1063/1.1302777.

[75] Bibring, J; *et al.* (2006). "Global mineralogical and aqueous mars history derived from OMEGA/Mars Express data". *Science* **312** (5772): 400–4. Bibcode 2006Sci...312..400B. doi:10.1126/science.1122659. PMID 16627738.

[76] Malik, Tariq (2005-03-08). "Hunt for Mars life should go underground" (http://www.msnbc.msn.com/id/7129347/). The Brown University News Bureau. . Retrieved September 4, 2006.

[77] Scott Turner (1998-03-02). "Detailed Images From Europa Point To Slush Below Surface" (http://www2.jpl.nasa.gov/galileo/news8.html). The Brown University News Bureau. . Retrieved September 28, 2006.

## External links

- Nature Journal – Weekly journal of science (http://www.nature.com/nature/index.html).
- The WILD Foundation (http://www.wild.org/).
- BBC – Science and Nature (http://www.bbc.co.uk/sn/).
- European Commission – Nature and Biodiversity homepage (http://ec.europa.eu/environment/nature/index_en.htm).
- The Nature Conservancy (http://www.nature.org/)
- Science.gov – Environment & Environmental Quality (http://www.science.gov/browse/w_123.htm).
- NatureWatch – Wiki for documenting biodiversity (german) (http://www.naturewatch.wikia.com/wiki/NatureWatch_Wiki)
- Ministry of Environment Republic of Korea – Nature (http://eng.me.go.kr/docs/sub2/policy_view.html?topmenu=C&cat=240&class=13)

# Cosmos

In the general sense, a **cosmos** is an orderly or harmonious system. It originates from the Greek term κόσμος (*kosmos*), meaning "order" or "ornament"[1] and is antithetical to the concept of chaos. Today, the word is generally used as a synonym of the word Universe (considered in its orderly aspect). The word cosmos originates from the same root. In many Slavic languages such as Russian and Bulgarian, the word *Космос* cosmos means also the "outer space". In Mandarin Chinese, cosmos is translated as *yuzhou*, which literally translated means space-time (*yu* = space + *zhou* = time).

The Ancient and Medieval cosmos as depicted in Peter Apian's *Cosmographia* (Antwerp, 1539).

## Philosophy

Pythagoras is said to have been the first philosopher to apply the term *cosmos* to the Universe, perhaps referring to the starry firmament.

The second largest extent of the Universe so far

Russian cosmism is a cosmocentric philosophical and cultural movement that emerged in Russia in the early 20th century.

Cosmicism is a philosophical position that mankind is an insignificant aspect of a universe at best indifferent and perhaps hostile. This philosophy, explored by writers such as H.P. Lovecraft (who some say is the original proponent of the philosophy) and later writers who actually represented the beliefs in books such as *Hitchhiker's Guide to the Galaxy*.

## Theology

In theology, the term can be used to denote the created Universe, not including the creator. In Christian theology, the word was also used synonymously with *aion* to refer to "worldly life" or "this world" as opposed to the afterlife or World to Come.

The cosmos as originated by Pythagoras is parallel to the Zoroastrian term *aša*, the concept of a divine order, or divinely ordered creation.

## Cosmology

Cosmology is the study of the cosmos in several of the above meanings, depending on context. All cosmologies have in common an attempt to understand the implicit order within the whole of being. In this way, most religions and philosophical systems have a cosmology.

Flammarion engraving, Paris 1888

Image of distribution of the cosmic microwave background radiation 700,000 years after the Big Bang, generally assumed to have occurred about 13,700,000,000 years ago.

In physical cosmology, the term *cosmos* is often used in a technical way, referring to a particular space-time continuum within the (postulated) multiverse. Our particular cosmos is generally capitalized as the *Cosmos*.

## Integral Philosophy

The philosopher Ken Wilber uses the term *kosmos* to refer to all of manifest existence, including various realms of consciousness. The term *kosmos* so used distinguishes a nondual Universe (which, in his view, includes both noetic and physical aspects) from the strictly physical Universe that is the concern of the traditional sciences. Wilber's nephew (Cosmo Iacavazzi, fullback at Princeton University) is said to have been named after the scientific term.

## Ancient Greek conception of the cosmos

The Ancient Greek natural philosopher Archimedes in his essay The Sand Reckoner, estimated the diameter of the cosmos to be the equivalent in stadia of what we call two light years.

## Age and size of the cosmos

According to current scientific theory, the cosmos began 13.7 billion years ago short scale in the Big Bang. The current diameter of the observable cosmos is thought to be about 93 billion light years.

The diameter of the entire cosmos is unknown. However, according to Alan Guth's inflation theory, the actual size of the cosmos is at least fifteen orders of magnitude larger than the observable universe. This means that if the inflation theory is correct, the 93 billion light year diameter of the observable universe is approximately as much smaller than the diameter of the entire universe as the diameter of a helium atom is compared to the diameter of the Sun. This is equivalent to a minimum diameter of the entire cosmos of $10^{26}$ light years (100 septillion light years short scale).

## See also

- Anthropic principle
- Astronomy
- Cosmic Logos
- Cosmic View
- Cosmic Zoom
- Cosmology
- Cosmogram
- Cosmography
- *Cosmos: A Personal Voyage* (Carl Sagan's television documentary series)
- Fate of the universe
- Lambda-CDM model
- Large scale structure of the cosmos
- Macrocosm and microcosm
- Megaverse
- Multiverse
- Observable universe
- Omega point (de Chardin)
- Omega point (Tipler)
- Omniverse
- Timeline of the Big Bang

## References

[1] κόσμος (http://www.perseus.tufts.edu/hopper/text?doc=Perseus:text:1999.04.0057:entry=ko/smos), Henry George Liddell, Robert Scott, *A Greek-English Lexicon*, on Perseus

## External links

- Cosmos – an Illustrated Dimensional Journey from microcosmos to macrocosmos (http://www.shekpvar.net/~dna/Publications/Cosmos/cosmos.html) – from Digital Nature Agency
- JPL Spitzer telescope photos of macrocosmos (http://www.spitzer.caltech.edu/search/image_set/20?by_subject=star&tabs=hidden)
- Macrocosm and Microcosm (http://etext.lib.virginia.edu/cgi-local/DHI/dhiana.cgi?id=dv3-16), in *Dictionary of the History of Ideas*
- Encyclopedia of Cosmos (http://www.eocosmos.org/)
- merriam-webster.com/dictionary/cosmic (http://www.merriam-webster.com/dictionary/cosmic)

# Universe

The **universe** is commonly defined as the totality of everything that exists,[1] including all matter and energy, the planets, stars, galaxies, and the contents of intergalactic space.[2] [3] Definitions and usage vary and similar terms include the *cosmos*, the *world* and *nature*. Scientific observation of earlier stages in the development of the universe, which can be seen at great distances, suggests that the universe has been governed by the same physical laws and constants throughout most of its extent and history. There are various multiverse theories, in which physicists have suggested that our universe is one among many universes that likewise exist.[4] [5]

## History

Throughout recorded history, several cosmologies and cosmogonies have been proposed to account for observations of the universe. The earliest quantitative geocentric models were developed by the ancient Greek philosophers. Over the centuries, more precise observations and improved theories of gravity led to Copernicus's heliocentric model and the Newtonian model of the Solar System, respectively. Further improvements in astronomy led to the realization that the Solar System is embedded in a galaxy composed of billions of stars, the Milky Way, and that other galaxies exist outside it, as far as astronomical instruments can reach. Careful studies of the distribution of these galaxies and their spectral lines have led to much of modern cosmology. Discovery of the red shift and cosmic microwave background radiation revealed that the universe is expanding and apparently had a beginning.

According to the prevailing scientific model of the universe, known as the Big Bang, the universe expanded from an extremely hot, dense phase called the Planck epoch, in which all the matter and energy of the observable universe was concentrated. Since the Planck epoch, the universe has been expanding to its present form, possibly with a brief period (less than $10^{-32}$ seconds) of cosmic inflation. Several independent experimental measurements support this theoretical expansion and, more generally, the Big Bang theory. Recent observations indicate that this expansion is accelerating because of dark energy, and that most of the matter in the universe may be in a form which cannot be detected by present instruments, and so is not accounted for in the present models of the universe; this has been named dark matter. The imprecision of current observations has hindered predictions of the ultimate fate of the universe.

This high-resolution image of the Hubble Ultra-Deep Field shows a diverse range of galaxies, each consisting of billions of stars. The equivalent area of sky that the picture occupies is shown as a red box in the lower left corner. The smallest, reddest galaxies, about 100, are some of the most distant galaxies to have been imaged by an optical telescope, existing at the time shortly after the Big Bang.

Current interpretations of astronomical observations indicate that the age of the universe is 13.75 ± 0.17 billion years,[6] and that the diameter of the observable universe is at least 93 billion light years or $8.80 \times 10^{26}$ metres.[7] According to general relativity, space can expand faster than the speed of light, although we can view only a small portion of the universe due to the limitation imposed by light speed. Since we cannot observe space beyond the limitations of light (or any electromagnetic radiation), it is uncertain whether the size of the universe is finite or infinite.

## Etymology, synonyms and definitions

The word *universe* derives from the Old French word *Univers*, which in turn derives from the Latin word *universum*.[8] The Latin word was used by Cicero and later Latin authors in many of the same senses as the modern English word is used.[9] The Latin word derives from the poetic contraction *Unvorsum* — first used by Lucretius in Book IV (line 262) of his *De rerum natura* (*On the Nature of Things*) — which connects *un, uni* (the combining form of *unus', or "one") with* vorsum, versum *(a noun made from the perfect passive participle of* vertere, *meaning "something rotated, rolled, changed").*[9]

An alternative interpretation of *unvorsum* is "everything rotated as one" or "everything rotated by one". In this sense, it may be considered a translation of an earlier Greek word for the universe, περιφορά, (periforá, "circumambulation"), originally used to describe a course of a meal, the food being carried around the circle of dinner guests.[10] This Greek word refers to celestial spheres, an early Greek model of the universe. Regarding Plato's Metaphor of the sun, Aristotle suggests that the rotation of the sphere of fixed stars inspired by the prime mover, motivates, in turn, terrestrial change via the Sun. Careful astronomical and physical measurements (such as the Foucault pendulum) are required to prove the Earth rotates on its axis.

A term for "universe" in ancient Greece was τὸ πᾶν (tó pán, The All, Pan (mythology)). Related terms were matter, (τὸ ὅλον, tó ólon, see also Hyle, lit. wood) and place (τὸ κενόν, tó kenón).[11] [12] Other synonyms for the universe among the ancient Greek philosophers included κόσμος (cosmos) and φύσις (meaning Nature, from which we derive the word physics).[13] The same synonyms are found in Latin authors (*totum*, *mundus*, *natura*)[14] and survive in modern languages, e.g., the German words *Das All*, *Weltall*, and *Natur* for universe. The same synonyms are found in English, such as everything (as in the theory of everything), the cosmos (as in cosmology), the world (as in the many-worlds hypothesis), and Nature (as in natural laws or natural philosophy).[15]

## Broadest definition: reality and probability

The broadest definition of the universe can be found in *De divisione naturae* by the medieval philosopher and theologian Johannes Scotus Eriugena, who defined it as simply everything: everything that is created and everything that is not created. In the path integral formulation of Richard Feynman,[16] the probability amplitudes for the various outcomes of an experiment given a perfectly defined initial state of the system are determined by summing over all possible histories (paths) by which the system could progress from the initial to final state.

## Definition as reality

More customarily, the universe is defined as everything that exists, (has existed, and will exist) . According to our current understanding, the universe consists of three principles: spacetime, forms of energy, including momentum and matter, and the physical laws that relate them.

## Definition as connected space-time

It is possible to conceive of disconnected space-times, each existing but unable to interact with one another. An easily visualized metaphor is a group of separate soap bubbles, in which observers living on one soap bubble cannot interact with those on other soap bubbles, even in principle. According to one common terminology, each "soap bubble" of space-time is denoted as a universe, whereas our particular space-time is denoted as *the universe*, just as we call our moon *the Moon*. The entire collection of these separate space-times is denoted as the multiverse.[17] In principle, the other unconnected universes may have different dimensionalities and topologies of space-time, different forms of matter and energy, and different physical laws and physical constants, although such possibilities are currently speculative.

## Definition as observable reality

According to a still-more-restrictive definition, the universe is everything within our connected space-time that could have a chance to interact with us and vice versa. According to the general theory of relativity, some regions of space may never interact with ours even in the lifetime of the universe, due to the finite speed of light and the ongoing expansion of space. For example, radio messages sent from Earth may never reach some regions of space, even if the universe would live forever; space may expand faster than light can traverse it. It is worth emphasizing that those distant regions of space are taken to exist and be part of reality as much as we are; yet we can never interact with them. The spatial region within which we can affect and be affected is denoted as the observable universe. Strictly speaking, the observable universe depends on the location of the observer. By traveling, an observer can come into contact with a greater region of space-time than an observer who remains still, so that the observable universe for the former is larger than for the latter. Nevertheless, even the most rapid traveler will not be able to interact with all of space. Typically, the observable universe is taken to mean the universe observable from our vantage point in the Milky Way Galaxy.

## Size, age, contents, structure, and laws

The universe is immensely large and possibly infinite in volume. The region visible from Earth (the observable universe) is a sphere with a radius of about 46 billion light years,[18] based on where the expansion of space has taken the most distant objects observed. For comparison, the diameter of a typical galaxy is only 30,000 light-years, and the typical distance between two neighboring galaxies is only 3 million light-years.[19] As an example, our Milky Way Galaxy is roughly 100,000 light years in diameter,[20] and our nearest sister galaxy, the Andromeda Galaxy, is located roughly 2.5 million light years away.[21] There are probably more than 100 billion ($10^{11}$) galaxies in the observable universe.[22] Typical galaxies range from dwarfs with as few as ten million[23] ($10^7$) stars up to giants with one trillion[24] ($10^{12}$) stars, all orbiting the galaxy's center of mass. A 2010 study by astronomers estimated that the observable universe contains 300 sextillion ($3\times10^{23}$) stars.[25]

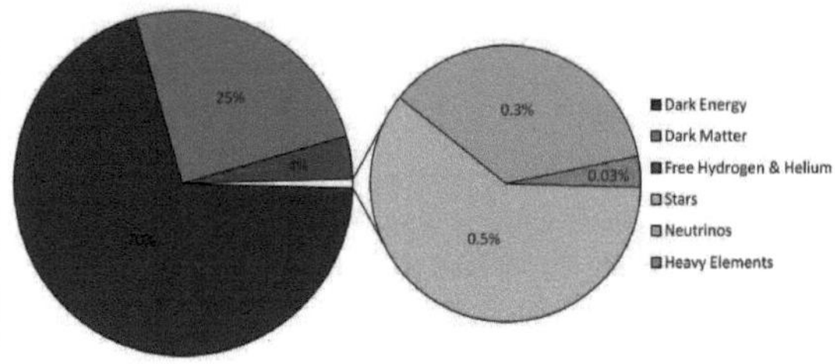


The universe is believed to be mostly composed of dark energy and dark matter, both of which are poorly understood at present. Less than 5% of the universe is ordinary matter, a relatively small contribution.

The observable matter is spread homogeneously (*uniformly*) throughout the universe, when averaged over distances longer than 300 million light-years.[26] However, on smaller length-scales, matter is observed to form "clumps", i.e., to cluster hierarchically; many atoms are condensed into stars, most stars into galaxies, most galaxies into clusters, superclusters and, finally, the largest-scale structures such as the Great Wall of galaxies. The observable matter of the universe is also spread *isotropically*, meaning that no direction of observation seems different from any other; each region of the sky has roughly the same content.[27] The universe is also bathed in a highly isotropic microwave radiation that corresponds to a thermal equilibrium blackbody spectrum of roughly 2.725-kelvins.[28] The hypothesis that the large-scale universe is homogeneous and isotropic is known as the cosmological principle,[29] which is supported by astronomical observations.

The present overall density of the universe is very low, roughly $9.9 \times 10^{-30}$ grams per cubic centimetre. This mass-energy appears to consist of 73% dark energy, 23% cold dark matter and 4% ordinary matter. Thus the density of atoms is on the order of a single hydrogen atom for every four cubic meters of volume.[30] The properties of dark energy and dark matter are largely unknown. Dark matter gravitates as ordinary matter, and thus works to slow the expansion of the universe; by contrast, dark energy accelerates its expansion.

The most precise estimate of the universe's age is 13.72±0.12 billion years old, based on observations of the cosmic microwave background radiation.[31] Independent estimates (based on measurements such as radioactive dating) agree at 13–15 billion years.[32] The universe has not been the same at all times in its history; for example, the relative populations of quasars and galaxies have changed and space itself appears to have expanded. This expansion accounts for how Earth-bound scientists can observe the light from a galaxy 30 billion light years away, even if that light has traveled for only 13 billion years; the very space between them has expanded. This expansion is consistent with the observation that the light from distant galaxies has been redshifted; the photons emitted have been stretched to longer wavelengths and lower frequency during their journey. The rate of this spatial expansion is accelerating, based on studies of Type Ia supernovae and corroborated by other data.

The relative fractions of different chemical elements — particularly the lightest atoms such as hydrogen, deuterium and helium — seem to be identical throughout the universe and throughout its observable history.[33] The universe seems to have much more matter than antimatter, an asymmetry possibly related to the observations of CP violation.[34] The universe appears to have no net electric charge, and therefore gravity appears to be the dominant interaction on cosmological length scales. The universe also appears to have neither net momentum nor angular momentum. The absence of net charge and momentum would follow from accepted physical laws (Gauss's law and the non-divergence of the stress-energy-momentum pseudotensor, respectively), if the universe were finite.[35]

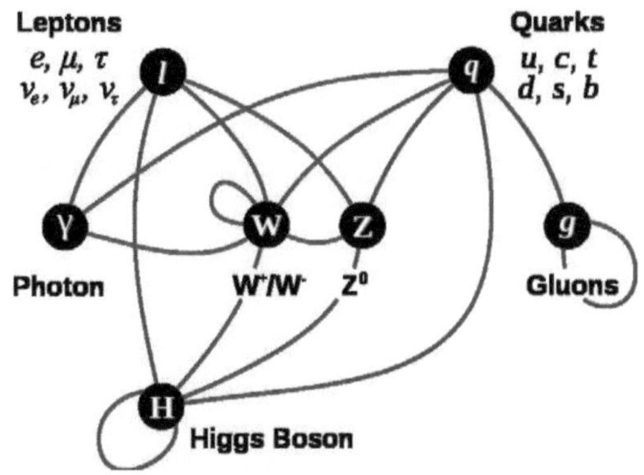


The elementary particles from which the universe is constructed. Six leptons and six quarks comprise most of the matter; for example, the protons and neutrons of atomic nuclei are composed of quarks, and the ubiquitous electron is a lepton. These particles interact via the gauge bosons shown in the middle row, each corresponding to a particular type of gauge symmetry. The Higgs boson (as yet unobserved) is believed to confer mass on the particles with which it is connected. The graviton, a supposed gauge boson for gravity, is not shown.

The universe appears to have a smooth space-time continuum consisting of three spatial dimensions and one temporal (time) dimension. On the average, space is observed to be very nearly flat (close to zero curvature), meaning that Euclidean geometry is experimentally true with high accuracy throughout most of the Universe.[36] Spacetime also appears to have a simply connected topology, at least on the length-scale of the observable universe. However, present observations cannot exclude the possibilities that the universe has more dimensions and that its spacetime may have a multiply connected global topology, in analogy with the cylindrical or toroidal topologies of two-dimensional spaces.[37]

The universe appears to behave in a manner that regularly follows a set of physical laws and physical constants.[38] According to the prevailing Standard Model of physics, all matter is composed of three generations of leptons and quarks, both of which are fermions. These elementary particles interact via at most three fundamental interactions: the electroweak interaction which includes electromagnetism and the weak nuclear force; the strong nuclear force described by quantum chromodynamics; and gravity, which is best described at present by general relativity. The first two interactions can be described by renormalized quantum field theory, and are mediated by gauge bosons that correspond to a particular type of gauge symmetry. A renormalized quantum field theory of general relativity has not yet been achieved, although various forms of string theory seem promising. The theory of special relativity is believed to hold throughout the universe, provided that the spatial and temporal length scales are sufficiently short; otherwise, the more general theory of general relativity must be applied. There is no explanation for the particular values that physical constants appear to have throughout our universe, such as Planck's constant $h$ or the gravitational constant $G$. Several conservation laws have been identified, such as the conservation of charge, momentum, angular momentum and energy; in many cases, these conservation laws can be related to symmetries or mathematical identities.

## Fine tuning

It appears that many of the properties of the universe have special values in the sense that a universe where these properties only differ slightly would not be able to support intelligent life.[39] [40] Not all scientists agree that this fine-tuning exists.[41] [42] In particular, it is not known under what conditions intelligent life could form and what form or shape that would take. A relevant observation in this discussion is that for an observer to exist to observe fine-tuning, the universe must be able to support intelligent life. As such the conditional probability of observing a universe that is fine-tuned to support intelligent life is 1. This observation is known as the anthropic principle and is particularly relevant if the creation of the universe was probabilistic or if multiple universes with a variety of properties exist (see below).

## Historical models

Many models of the cosmos (cosmologies) and its origin (cosmogonies) have been proposed, based on the then-available data and conceptions of the universe. Historically, cosmologies and cosmogonies were based on narratives of gods acting in various ways. Theories of an impersonal universe governed by physical laws were first proposed by the Greeks and Indians. Over the centuries, improvements in astronomical observations and theories of motion and gravitation led to ever more accurate descriptions of the universe. The modern era of cosmology began with Albert Einstein's 1915 general theory of relativity, which made it possible to quantitatively predict the origin, evolution, and conclusion of the universe as a whole. Most modern, accepted theories of cosmology are based on general relativity and, more specifically, the predicted Big Bang; however, still more careful measurements are required to determine which theory is correct.

### Creation

Many cultures have stories describing the origin of the world, which may be roughly grouped into common types. In one type of story, the world is born from a world egg; such stories include the Finnish epic poem *Kalevala*, the Chinese story of Pangu or the Indian Brahmanda Purana. In related stories, the creation idea is caused by a single entity emanating or producing something by him- or herself, as in the Tibetan Buddhism concept of Adi-Buddha, the ancient Greek story of Gaia (Mother Earth), the Aztec goddess Coatlicue myth, the ancient Egyptian god Atum story, or the Genesis creation narrative. In another type of story, the world is created from the union of male and female deities, as in the Maori story of Rangi and Papa. In other stories, the universe is created by crafting it from pre-existing materials, such as the corpse of a dead god — as from Tiamat in the Babylonian epic Enuma Elish or from the giant Ymir in Norse mythology – or from chaotic materials, as in Izanagi and Izanami in Japanese mythology. In other stories, the universe emanates from fundamental principles, such as Brahman and Prakrti, or the yin and yang of the Tao.

### Philosophical models

Further information: Cosmology

From the 6th century BCE, the pre-Socratic Greek philosophers developed the earliest known philosophical models of the universe. The earliest Greek philosophers noted that appearances can be deceiving, and sought to understand the underlying reality behind the appearances. In particular, they noted the ability of matter to change forms (e.g., ice to water to steam) and several philosophers proposed that all the apparently different materials of the world are different forms of a single primordial material, or arche. The first to do so was Thales, who proposed this material is Water. Thales' student, Anaximander, proposed that everything came from the limitless apeiron. Anaximenes proposed Air on account of its perceived attractive and repulsive qualities that cause the arche to condense or dissociate into different forms. Anaxagoras, proposed the principle of Nous (Mind). Heraclitus proposed fire (and spoke of logos). Empedocles proposed the elements: earth, water, air and fire. His four element theory became very popular. Like Pythagoras, Plato believed that all things were composed of number, with the Empedocles' elements taking the form of the Platonic solids. Democritus, and later philosophers—most notably Leucippus—proposed that the universe was composed of indivisible atoms moving through void (vacuum). Aristotle did not believe that was feasible because air, like water, offers resistance to motion. Air will immediately rush in to fill a void, and moreover, without resistance, it would do so indefinitely fast.

Although Heraclitus argued for eternal change, his quasi-contemporary Parmenides made the radical suggestion that all change is an illusion, that the true underlying reality is eternally unchanging and of a single nature. Parmenides denoted this reality as τὸ ἐν (The One). Parmenides' theory seemed implausible to many Greeks, but his student Zeno of Elea challenged them with several famous paradoxes. Aristotle responded to these paradoxes by developing the notion of a potential countable infinity, as well as the infinitely divisible continuum. Unlike the eternal and unchanging cycles of time, he believed the world was bounded by the celestial spheres, and thus magnitude was only finitely multiplicative.

The Indian philosopher Kanada, founder of the Vaisheshika school, developed a theory of atomism and proposed that light and heat were varieties of the same substance.[43] In the 5th century AD, the Buddhist atomist philosopher Dignāga proposed atoms to be point-sized, durationless, and made of energy. They denied the existence of substantial matter and proposed that movement consisted of momentary flashes of a stream of energy.[44]

The theory of temporal finitism was inspired by the doctrine of Creation shared by the three Abrahamic religions: Judaism, Christianity and Islam. The Christian philosopher, John Philoponus, presented the philosophical arguments against the ancient Greek notion of an infinite past and future. Philoponus' arguments against an infinite past were used by the early Muslim philosopher, Al-Kindi (Alkindus); the Jewish philosopher, Saadia Gaon (Saadia ben Joseph); and the Muslim theologian, Al-Ghazali (Algazel). Borrowing from Aristotle's *Physics* and *Metaphysics*, they employed two logical arguments against an infinite past, the first being the "argument from the impossibility of the existence of an actual infinite", which states:[45]

> "An actual infinite cannot exist."
>
> "An infinite temporal regress of events is an actual infinite."
>
> " An infinite temporal regress of events cannot exist."

The second argument, the "argument from the impossibility of completing an actual infinite by successive addition", states:[45]

> "An actual infinite cannot be completed by successive addition."
>
> "The temporal series of past events has been completed by successive addition."
>
> " The temporal series of past events cannot be an actual infinite."

Both arguments were adopted by Christian philosophers and theologians, and the second argument in particular became more famous after it was adopted by Immanuel Kant in his thesis of the first antinomy concerning time.[45]

## Astronomical models

Astronomical models of the universe were proposed soon after astronomy began with the Babylonian astronomers, who viewed the universe as a flat disk floating in the ocean, and this forms the premise for early Greek maps like those of Anaximander and Hecataeus of Miletus.

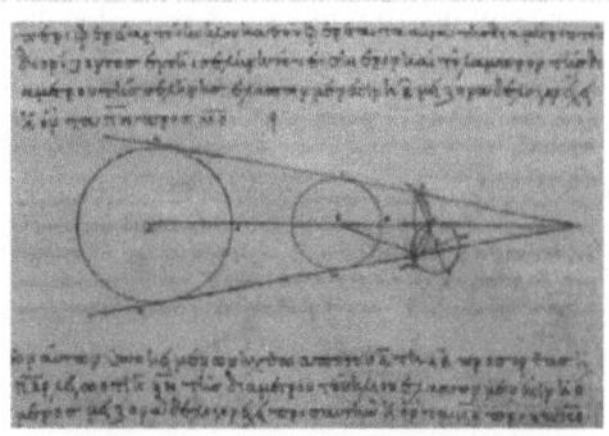

Aristarchus's 3rd century BCE calculations on the relative sizes of from left the Sun, Earth and Moon, from a 10th century AD Greek copy

Later Greek philosophers, observing the motions of the heavenly bodies, were concerned with developing models of the universe based more profoundly on empirical evidence. The first coherent model was proposed by Eudoxus of Cnidos. According to Aristotle's physical interpretation of the model, celestial spheres eternally rotate with uniform motion around a stationary Earth. Normal matter, is entirely contained within the terrestrial sphere. This model was also refined by Callippus and after concentric spheres were abandoned, it was brought into nearly perfect agreement with astronomical observations by Ptolemy. The success of such a model is largely due to the mathematical fact that any function (such as the position of a planet) can be decomposed into a set of circular functions (the Fourier modes). Other Greek scientists, such as the Pythagorean philosopher Philolaus postulated that at the center of the universe was a "central fire" around which the Earth, Sun, Moon and Planets revolved in uniform circular motion.[46] The Greek astronomer Aristarchus of Samos was the first known individual to propose a heliocentric model of the universe. Though the original text has been lost, a reference in Archimedes' book The Sand Reckoner describes Aristarchus' heliocentric theory. Archimedes wrote: (translated into English)

> You King Gelon are aware the 'universe' is the name given by most astronomers to the sphere the center of which is the center of the Earth, while its radius is equal to the straight line between the center of the Sun and the center of the Earth. This is the common account as you have heard from astronomers. But Aristarchus has brought out a book consisting of certain hypotheses, wherein it appears, as a consequence of the assumptions made, that the universe is many times greater than the 'universe' just mentioned. His hypotheses are that the fixed stars and the Sun remain unmoved, that the Earth revolves about the Sun on the circumference of a circle, the Sun lying in the middle of the orbit, and that the sphere of fixed stars, situated about the same center as the Sun, is so great that the circle in which he supposes the Earth to revolve bears such a proportion to the distance of the fixed stars as the center of the sphere bears to its surface.

Aristarchus thus believed the stars to be very far away, and saw this as the reason why there was no visible parallax, that is, an observed movement of the stars relative to each other as the Earth moved around the Sun. The stars are in fact much farther away than the distance that was generally assumed in ancient times, which is why stellar parallax is only detectable with telescopes. The geocentric model, consistent with planetary parallax, was assumed to be an explanation for the unobservability of the parallel phenomenon, stellar parallax. The rejection of the heliocentric view was apparently quite strong, as the following passage from Plutarch suggests (On the Apparent Face in the Orb of the Moon):

> Cleanthes [a contemporary of Aristarchus and head of the Stoics] thought it was the duty of the Greeks to indict Aristarchus of Samos on the charge of impiety for putting in motion the Hearth of the universe [i.e. the earth], . . . supposing the heaven to remain at rest and the earth to revolve in an oblique circle, while it rotates, at the same time, about its own axis. [1]

The only other astronomer from antiquity known by name who supported Aristarchus' heliocentric model was Seleucus of Seleucia, a Hellenistic astronomer who lived a century after Aristarchus.[47] [48] [49] According to Plutarch, Seleucus was the first to prove the heliocentric system through reasoning, but it is not known what arguments he used. Seleucus' arguments for a heliocentric theory were probably related to the phenomenon of tides.[50] According to Strabo (1.1.9), Seleucus was the first to state that the tides are due to the attraction of the Moon, and that the height of the tides depends on the Moon's position relative to the Sun.[51] Alternatively, he may have proved the heliocentric theory by determining the constants of a geometric model for the heliocentric theory and by developing methods to compute planetary positions using this model, like what Nicolaus Copernicus later did in the 16th century.[52] During the Middle Ages, heliocentric models may have also been proposed by the Indian astronomer, Aryabhata,[53] and by the Persian astronomers, Albumasar[54] and Al-Sijzi.[55]

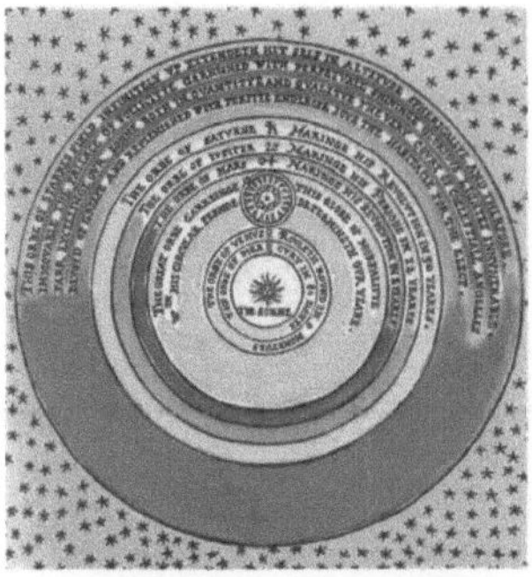

Model of the Copernican universe by Thomas Digges in 1576, with the amendment that the stars are no longer confined to a sphere, but spread uniformly throughout the space surrounding the planets.

The Aristotelian model was accepted in the Western world for roughly two millennia, until Copernicus revived Aristarchus' theory that the astronomical data could be explained more plausibly if the earth rotated on its axis and if the sun were placed at the center of the universe.

> In the center rests the sun. For who would place this lamp of a very beautiful temple in another or better place than this wherefrom it can illuminate everything at the same time?
>
> —Nicolaus Copernicus, in Chapter 10, Book 1 of De Revolutionibus Orbium Coelestrum (1543)

As noted by Copernicus himself, the suggestion that the Earth rotates was very old, dating at least to Philolaus (c. 450 BC), Heraclides Ponticus (c. 350 BC) and Ecphantus the Pythagorean. Roughly a century before Copernicus, Christian scholar Nicholas of Cusa also proposed that the Earth rotates on its axis in his book, *On Learned Ignorance* (1440).[56] Aryabhata (476–550), Brahmagupta (598–668), Albumasar and Al-Sijzi, also proposed that the Earth rotates on its axis. The first empirical evidence for the Earth's rotation on its axis, using the phenomenon of comets, was given by Tusi (1201–1274) and Ali Qushji (1403–1474).

Johannes Kepler published the *Rudolphine Tables* containing a star catalog and planetary tables using Tycho Brahe's measurements.

This cosmology was accepted by Isaac Newton, Christiaan Huygens and later scientists.[57] Edmund Halley (1720)[58] and Jean-Philippe de Cheseaux (1744)[59] noted independently that the assumption of an infinite space filled uniformly with stars would lead to the prediction that the nighttime sky would be as bright as the sun itself; this became known as Olbers' paradox in the 19th century.[60] Newton believed that an infinite space uniformly filled with matter would cause infinite forces and instabilities causing the matter to be crushed inwards under its own gravity.[57] This instability was clarified in 1902 by the Jeans instability criterion.[61] One solution to these paradoxes is the Charlier universe, in which the matter is arranged hierarchically (systems of orbiting bodies that are themselves orbiting in a larger system, *ad infinitum*) in a fractal way such that the universe has a negligibly small overall density; such a cosmological model had also been proposed earlier in 1761 by Johann Heinrich Lambert.[62] A significant astronomical advance of the 18th century was the realization by Thomas Wright, Immanuel Kant and others of nebulae.[63]

The modern era of physical cosmology began in 1917, when Albert Einstein first applied his general theory of relativity to model the structure and dynamics of the universe.[64]

# Theoretical models

Of the four fundamental interactions, gravitation is dominant at cosmological length scales; that is, the other three forces are believed to play a negligible role in determining structures at the level of planets, stars, galaxies and larger-scale structures. Since all matter and energy gravitate, gravity's effects are cumulative; by contrast, the effects of positive and negative charges tend to cancel one another, making electromagnetism relatively insignificant on cosmological length scales. The remaining two interactions, the weak and strong nuclear forces, decline very rapidly with distance; their effects are confined mainly to sub-atomic length scales.

High-precision test of general relativity by the Cassini space probe (artist's impression): radio signals sent between the Earth and the probe (green wave) are delayed by the warping of space and time (blue lines) due to the Sun's mass.

## General theory of relativity

Given gravitation's predominance in shaping cosmological structures, accurate predictions of the universe's past and future require an accurate theory of gravitation. The best theory available is Albert Einstein's general theory of relativity, which has passed all experimental tests hitherto. However, since rigorous experiments have not been carried out on cosmological length scales, general relativity could conceivably be inaccurate. Nevertheless, its cosmological predictions appear to be consistent with observations, so there is no compelling reason to adopt another theory.

General relativity provides a set of ten nonlinear partial differential equations for the spacetime metric (Einstein's field equations) that must be solved from the distribution of mass-energy and momentum throughout the universe. Since these are unknown in exact detail, cosmological models have been based on the cosmological principle, which states that the universe is homogeneous and isotropic. In effect, this principle asserts that the gravitational effects of the various galaxies making up the universe are equivalent to those of a fine dust distributed uniformly throughout the universe with the same average density. The assumption of a uniform dust makes it easy to solve Einstein's field equations and predict the past and future of the universe on cosmological time scales.

Einstein's field equations include a cosmological constant ($\Lambda$),[64] [65] that corresponds to an energy density of empty space.[66] Depending on its sign, the cosmological constant can either slow (negative $\Lambda$) or accelerate (positive $\Lambda$) the expansion of the universe. Although many scientists, including Einstein, had speculated that $\Lambda$ was zero,[67] recent astronomical observations of type Ia supernovae have detected a large amount of "dark energy" that is accelerating the universe's expansion.[68] Preliminary studies suggest that this dark energy corresponds to a positive $\Lambda$, although alternative theories cannot be ruled out as yet.[69] Russian physicist Zel'dovich suggested that $\Lambda$ is a measure of the zero-point energy associated with virtual particles of quantum field theory, a pervasive vacuum energy that exists everywhere, even in empty space.[70] Evidence for such zero-point energy is observed in the Casimir effect.

## Special relativity and space-time

The universe has at least three spatial and one temporal (time) dimension. It was long thought that the spatial and temporal dimensions were different in nature and independent of one another. However, according to the special theory of relativity, spatial and temporal separations are interconvertible (within limits) by changing one's motion.

To understand this interconversion, it is helpful to consider the analogous interconversion of spatial separations along the three spatial dimensions. Consider the two endpoints of a rod of length *L*. The length can be determined from the differences in the three coordinates Δx, Δy and Δz of the two endpoints in a given reference frame

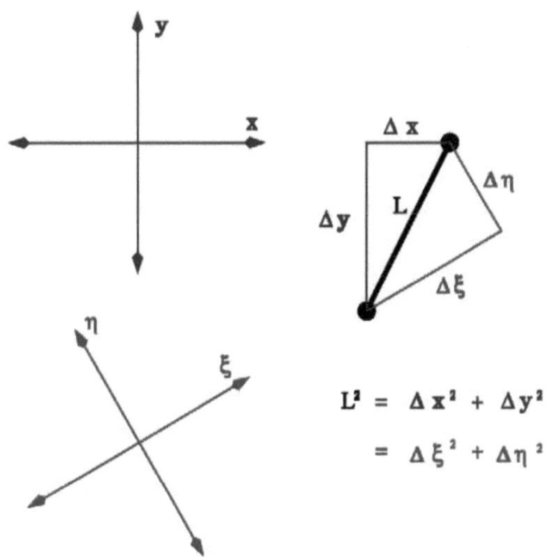


Only its length *L* is intrinsic to the rod (shown in black); coordinate differences between its endpoints (such as Δx, Δy or Δξ, Δη) depend on their frame of reference (depicted in blue and red, respectively).

using the Pythagorean theorem. In a rotated reference frame, the coordinate differences differ, but they give the same length

Thus, the coordinates differences (Δx, Δy, Δz) and (Δξ, Δη, Δζ) are not intrinsic to the rod, but merely reflect the reference frame used to describe it; by contrast, the length *L* is an intrinsic property of the rod. The coordinate differences can be changed without affecting the rod, by rotating one's reference frame.

The analogy in spacetime is called the interval between two events; an event is defined as a point in spacetime, a specific position in space and a specific moment in time. The spacetime interval between two events is given by

where *c* is the speed of light. According to special relativity, one can change a spatial and time separation ($L_1$, $\Delta t_1$) into another ($L_2$, $\Delta t_2$) by changing one's reference frame, as long as the change maintains the spacetime interval *s*. Such a change in reference frame corresponds to changing one's motion; in a moving frame, lengths and times are different from their counterparts in a stationary reference frame. The precise manner in which the coordinate and time differences change with motion is described by the Lorentz transformation.

## Solving Einstein's field equations

The distances between the spinning galaxies increase with time, but the distances between the stars within each galaxy stay roughly the same, due to their gravitational interactions. This animation illustrates a closed Friedmann universe with zero cosmological constant Λ; such a universe oscillates between a Big Bang and a Big Crunch.

In non-Cartesian (non-square) or curved coordinate systems, the Pythagorean theorem holds only on infinitesimal length scales and must be augmented with a more general metric tensor $g_{\mu\nu}$, which can vary from place to place and which describes the local geometry in the particular coordinate system. However, assuming the cosmological principle that the universe is homogeneous and isotropic everywhere, every point in space is like every other point; hence, the metric tensor must be the same everywhere. That leads to a single form for the metric tensor, called the Friedmann–Lemaître–Robertson–Walker metric where $(r, \theta, \varphi)$ correspond to a spherical coordinate system. This metric has only two undetermined parameters: an overall length scale $R$ that can vary with time, and a curvature index $k$ that can be only 0, 1 or −1, corresponding to flat Euclidean geometry, or spaces of positive or negative curvature. In cosmology, solving for the history of the universe is done by calculating $R$ as a function of time, given $k$ and the value of the cosmological constant $\Lambda$, which is a (small) parameter in Einstein's field equations. The equation describing how $R$ varies with time is known as the Friedmann equation, after its inventor, Alexander Friedmann.[71]

The solutions for $R(t)$ depend on $k$ and $\Lambda$, but some qualitative features of such solutions are general. First and most importantly, the length scale $R$ of the universe can remain constant *only* if the universe is perfectly isotropic with positive curvature ($k$=1) and has one precise value of density everywhere, as first noted by Albert Einstein. However, this equilibrium is unstable and since the universe is known to be inhomogeneous on smaller scales, $R$ must change, according to general relativity. When $R$ changes, all the spatial distances in the universe change in tandem; there is an overall expansion or contraction of space itself. This accounts for the observation that galaxies appear to be flying apart; the space between them is stretching. The stretching of space also accounts for the apparent paradox that two galaxies can be 40 billion light years apart, although they started from the same point 13.7 billion years ago and never moved faster than the speed of light.

Second, all solutions suggest that there was a gravitational singularity in the past, when $R$ goes to zero and matter and energy became infinitely dense. It may seem that this conclusion is uncertain since it is based on the questionable assumptions of perfect homogeneity and isotropy (the cosmological principle) and that only the gravitational interaction is significant. However, the Penrose–Hawking singularity theorems show that a singularity should exist for very general conditions. Hence, according to Einstein's field equations, $R$ grew rapidly from an unimaginably hot, dense state that existed immediately following this singularity (when $R$ had a small, finite value); this is the essence of the Big Bang model of the universe. A common misconception is that the Big Bang model predicts that matter and energy exploded from a single point in space and time; that is false. Rather, space itself was created in the Big Bang and imbued with a fixed amount of energy and matter distributed uniformly throughout; as space expands (i.e., as $R(t)$ increases), the density of that matter and energy decreases.

| Space has no boundary – that is empirically more certain than any external observation. However, that does not imply that space is infinite...(translated, original German) |
|---|
| Bernhard Riemann (Habilitationsvortrag, 1854) |

Third, the curvature index $k$ determines the sign of the mean spatial curvature of spacetime averaged over length scales greater than a billion light years. If $k$=1, the curvature is positive and the universe has a finite volume. Such universes are often visualized as a three-dimensional sphere $S^3$ embedded in a four-dimensional space. Conversely, if $k$ is zero or negative, the universe *may* have infinite volume, depending on its overall topology. It may seem counter-intuitive that an infinite and yet infinitely dense universe could be created in a single instant at the Big Bang when $R$=0, but exactly that is predicted mathematically when $k$ does not equal 1. For comparison, an infinite plane has zero curvature but infinite area, whereas an infinite cylinder is finite in one direction and a torus is finite in both. A toroidal universe could behave like a normal universe with periodic boundary conditions, as seen in "wrap-around" video games such as *Asteroids*; a traveler crossing an outer "boundary" of space going *outwards* would reappear instantly at another point on the boundary moving *inwards*.

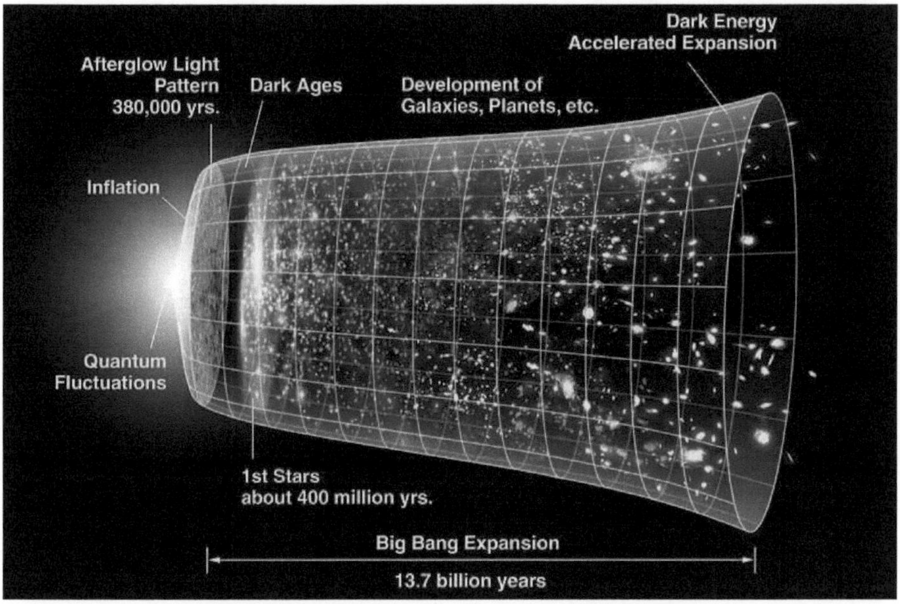


Prevailing model of the origin and expansion of spacetime and all that it contains.

The ultimate fate of the universe is still unknown, because it depends critically on the curvature index $k$ and the cosmological constant $\Lambda$. If the universe is sufficiently dense, $k$ equals +1, meaning that its average curvature throughout is positive and the universe will eventually recollapse in a Big Crunch, possibly starting a new universe in a Big Bounce. Conversely, if the universe is insufficiently dense, $k$ equals 0 or −1 and the universe will expand forever, cooling off and eventually becoming inhospitable for all life, as the stars die and all matter coalesces into black holes (the Big Freeze and the heat death of the universe). As noted above, recent data suggests that the expansion speed of the universe is not decreasing as originally expected, but increasing; if this continues indefinitely, the universe will eventually rip itself to shreds (the Big Rip). Experimentally, the universe has an overall density that is very close to the critical value between recollapse and eternal expansion; more careful astronomical observations are needed to resolve the question.

## Big Bang model

The prevailing Big Bang model accounts for many of the experimental observations described above, such as the correlation of distance and redshift of galaxies, the universal ratio of hydrogen:helium atoms, and the ubiquitous, isotropic microwave radiation background. As noted above, the redshift arises from the metric expansion of space; as the space itself expands, the wavelength of a photon traveling through space likewise increases, decreasing its energy. The longer a photon has been traveling, the more expansion it has undergone; hence, older photons from more distant galaxies are the most red-shifted. Determining the correlation between distance and redshift is an important problem in experimental physical cosmology.

Other experimental observations can be explained by combining the overall expansion of space with nuclear and atomic physics. As the universe expands, the energy density of the electromagnetic radiation decreases more quickly than does that of matter, since the energy of a photon decreases with its wavelength. Thus, although the energy density of the universe is now dominated by matter, it was once dominated by radiation; poetically speaking, all was light. As the universe expanded, its energy density decreased and it became cooler; as it did so, the elementary particles of matter could associate stably into ever larger combinations. Thus, in the early part of the matter-dominated era, stable protons and neutrons formed, which then associated into atomic nuclei. At this stage, the matter in the universe was mainly a hot, dense plasma of negative electrons, neutral neutrinos and positive nuclei. Nuclear reactions among the nuclei led to the present abundances of the lighter nuclei, particularly hydrogen, deuterium, and helium. Eventually, the electrons and nuclei combined to form stable atoms, which are transparent to most wavelengths of radiation; at this point, the radiation decoupled from the matter, forming the ubiquitous, isotropic background of microwave radiation observed today.

$$
\begin{array}{ll}
\mathrm{n}^0 \longrightarrow \mathrm{p}^+ + \mathrm{e}^- + \bar{\nu}_e & \mathrm{p}^+ + \mathrm{n}^0 \longrightarrow {}^2_1\mathrm{D} + \gamma \\
{}^2_1\mathrm{D} + \mathrm{p}^+ \longrightarrow {}^3_2\mathrm{He} + \gamma & {}^2_1\mathrm{D} + {}^2_1\mathrm{D} \longrightarrow {}^3_2\mathrm{He} + \mathrm{n}^0 \\
{}^2_1\mathrm{D} + {}^2_1\mathrm{D} \longrightarrow {}^3_1\mathrm{T} + \mathrm{p}^+ & {}^3_1\mathrm{T} + {}^2_1\mathrm{D} \longrightarrow {}^4_2\mathrm{He} + \mathrm{n}^0 \\
{}^3_2\mathrm{He} + {}^4_2\mathrm{He} \longrightarrow {}^7_3\mathrm{Li} + \gamma & {}^3_2\mathrm{He} + \mathrm{n}^0 \longrightarrow {}^3_1\mathrm{T} + \mathrm{p}^+ \\
{}^3_2\mathrm{He} + {}^2_1\mathrm{D} \longrightarrow {}^4_2\mathrm{He} + \mathrm{p}^+ & {}^3_2\mathrm{He} + {}^4_2\mathrm{He} \longrightarrow {}^7_4\mathrm{Be} + \gamma \\
{}^7_3\mathrm{Li} + \mathrm{p}^+ \longrightarrow {}^4_2\mathrm{He} + {}^4_2\mathrm{He} & {}^7_4\mathrm{Be} + \mathrm{n}^0 \longrightarrow {}^7_3\mathrm{Li} + \mathrm{p}^+
\end{array}
$$

Chief nuclear reactions responsible for the relative abundances of light atomic nuclei observed throughout the universe.

Other observations are not answered definitively by known physics. According to the prevailing theory, a slight imbalance of matter over antimatter was present in the universe's creation, or developed very shortly thereafter, possibly due to the CP violation that has been observed by particle physicists. Although the matter and antimatter mostly annihilated one another, producing photons, a small residue of matter survived, giving the present matter-dominated universe. Several lines of evidence also suggest that a rapid cosmic inflation of the universe occurred very early in its history (roughly $10^{-35}$ seconds after its creation). Recent observations also suggest that the cosmological constant ($\Lambda$) is not zero and that the net mass-energy content of the universe is dominated by a dark energy and dark matter that have not been characterized scientifically. They differ in their gravitational effects. Dark matter gravitates as ordinary matter does, and thus slows the expansion of the universe; by contrast, dark energy serves to accelerate the universe's expansion.

## Multiverse theory

Some speculative theories have proposed that this universe is but one of a set of disconnected universes, collectively denoted as the multiverse, challenging or enhancing more limited definitions of the universe.[17] [72] Scientific multiverse theories are distinct from concepts such as alternate planes of consciousness and simulated reality, although the idea of a larger universe is not new; for example, Bishop Étienne Tempier of Paris ruled in 1277 that God could create as many universes as he saw fit, a question that was being hotly debated by the French theologians.[73]

Depiction of a multiverse of seven "bubble" universes, which are separate spacetime continua, each having different physical laws, physical constants, and perhaps even different numbers of dimensions or topologies.

Max Tegmark developed a four part classification scheme for the different types of multiverses that scientists have suggested in various problem domains. An example of such a theory is the chaotic inflation model of the early universe.[74] Another is the many-worlds interpretation of quantum mechanics. Parallel worlds are generated in a manner similar to quantum superposition and decoherence, with all states of the wave function being realized in separate worlds. Effectively, the multiverse evolves as a universal wavefunction. If the big bang that created our multiverse created an ensemble of multiverses, the wave function of the ensemble would be entangled in this sense.

The least controversial category of multiverse in Tegmark's scheme is Level I, which describes distant space-time events "in our own universe". If space is infinite, or sufficiently large and uniform, identical instances of the history of Earth's entire Hubble volume occur every so often, simply by chance. Tegmark calculated our nearest so-called doppelgänger, is $10^{10}115$ meters away from us (a double exponential function larger than a googolplex).[75] [76] In principle, it would be impossible to scientifically verify an identical Hubble volume. However, it does follow as a fairly straightforward consequence from otherwise unrelated scientific observations and theories. Tegmark suggests that statistical analysis exploiting the anthropic principle provides an opportunity to test multiverse theories in some cases. Generally, science would consider a multiverse theory that posits neither a common point of causation, nor the possibility of interaction between universes, to be an ideal speculation.

## Shape of the universe

The shape or geometry of the universe includes both local geometry in the observable universe and global geometry, which we may or may not be able to measure. Shape can refer to curvature and topology. More formally, the subject in practice investigates which 3-manifold corresponds to the spatial section in comoving coordinates of the four-dimensional space-time of the universe. Cosmologists normally work with a given space-like slice of spacetime called the comoving coordinates. In terms of observation, the section of spacetime that can be observed is the backward light cone (points within the cosmic light horizon, given time to reach a given observer). If the observable universe is smaller than the entire universe (in some models it is many orders of magnitude smaller), one cannot determine the global structure by observation: one is limited to a small patch.

Among the Friedmann–Lemaître–Robertson–Walker (FLRW) models, the presently most popular shape of the Universe found to fit observational data according to cosmologists is the infinite flat model,[77] while other FLRW models include the Poincaré dodecahedral space[78] [79] and the Picard horn.[80] The data fit by these FLRW models of space especially include the Wilkinson Microwave Anisotropy Probe (WMAP) maps of cosmic background radiation. NASA released the first WMAP cosmic background radiation data in February 2003. In 2009 the Planck observatory was launched to observe the microwave background at higher resolution than WMAP, possibly providing more information on the shape of the Universe. The data should be released in late 2012.

## See also

- Cosmic latte
- Cosmology
- Dyson's eternal intelligence
- Esoteric cosmology
- False vacuum
- Final anthropic principle
- Fine-tuned Universe
- Hindu cycle of the universe
- Jain cosmology
- Kardashev scale
- *The Mysterious Universe* (book)
- Nucleocosmochronology
- Non-standard cosmology
- Observable universe

- Omega Point
- Omniverse
- Rare Earth hypothesis
- Vacuum genesis
- World view
- Zero-energy universe

## Notes and references


[1] *Webster's New World College Dictionary* (http://www.yourdictionary.com/universe). Wiley Publishing, Inc.. 2010. .

[2] *The American Heritage Dictionary of the English Language* (http://www.yourdictionary.com/universe) (4th ed.). Houghton Mifflin Harcourt Publishing Company. 2010. .

[3] *Cambridge Advanced Learner's Dictionary* (http://dictionary.cambridge.org/dictionary/british/universe). .

[4] multiverse (http://www.astronomy.pomona.edu/Projects/moderncosmo/Sean's mutliverse.html). Astronomy.pomona.edu. Retrieved on 2011-11-28.

[5] Palmer, Jason. (2011-08-03) BBC News – 'Multiverse' theory suggested by microwave background (http://www.bbc.co.uk/news/science-environment-14372387). Retrieved on 2011-11-28.

[6] Suyu, S. H.; Marshall, P. J.; Auger, M. W.; Hilbert, S.; Blandford, R. D.; Koopmans, L. V. E.; Fassnacht, C. D.; Treu, T. (2010). "Dissecting the gravitational lens B1608+656. II. Precision measurements of the Hubble constant, spatial curvature, and the dark energy equation of state". *The Astrophysical Journal* **711**: 201–221. Bibcode 2010ApJ...711..201S. doi:10.1088/0004-637X/711/1/201.

[7] Lineweaver, Charles; Davis, Tamara M. (2005). "Misconceptions about the Big Bang" (http://www.sciam.com/article.cfm?id=misconceptions-about-the-2005-03&page=5). Scientific American. . Retrieved 2008-11-06.

[8] *The Compact Edition of the Oxford English Dictionary*, volume II, Oxford: Oxford University Press, 1971, p.3518.

[9] Lewis, C. T. and Short, S *A Latin Dictionary*, Oxford University Press, ISBN 0-19-864201-6, pp. 1933, 1977–1978.

[10] Liddell and Scott, p. 1392.

[11] Liddell and Scott, pp. 1345–1346.

[12] Yonge, Charles Duke (1870). *An English-Greek lexicon*. New York: American Bok Company. p. 567.

[13] Liddell and Scott, pp. 985, 1964.

[14] Lewis and Short, pp. 1881–1882, 1175, 1189–1190.

[15] OED, pp. 909, 569, 3821–3822, 1900.

[16] Feynman RP, Hibbs AR (1965). *Quantum Physics and Path Integrals*. New York: McGraw–Hill. ISBN 0-07-020650-3.
Zinn Justin J (2004). *Path Integrals in Quantum Mechanics*. Oxford University Press. ISBN 0-19-856674-3. OCLC 212409192.

[17] Ellis, George F.R.; U. Kirchner, W.R. Stoeger (2004). "Multiverses and physical cosmology". *Monthly Notices of the Royal Astronomical Society* **347** (3): 921–936. arXiv:astro-ph/0305292. Bibcode 2004MNRAS.347..921E. doi:10.1111/j.1365-2966.2004.07261.x.

[18] Lineweaver, Charles; Tamara M. Davis (2005). "Misconceptions about the Big Bang" (http://space.mit.edu/~kcooksey/teaching/AY5/MisconceptionsabouttheBigBang_ScientificAmerican.pdf). Scientific American. . Retrieved 2007-03-05.

[19] Rindler (1977), p.196.

[20] Christian, Eric; Samar, Safi-Harb. "How large is the Milky Way?" (http://imagine.gsfc.nasa.gov/docs/ask_astro/answers/980317b.html). . Retrieved 2007-11-28.

[21] I. Ribas, C. Jordi, F. Vilardell, E.L. Fitzpatrick, R.W. Hilditch, F. Edward (2005). "First Determination of the Distance and Fundamental Properties of an Eclipsing Binary in the Andromeda Galaxy". *Astrophysical Journal* **635** (1): L37–L40. arXiv:astro-ph/0511045. Bibcode 2005ApJ...635L..37R. doi:10.1086/499161.
McConnachie, A. W.; Irwin, M. J.; Ferguson, A. M. N.; Ibata, R. A.; Lewis, G. F.; Tanvir, N. (2005). "Distances and metallicities for 17 Local Group galaxies". *Monthly Notices of the Royal Astronomical Society* **356** (4): 979–997. arXiv:astro-ph/0410489. Bibcode 2005MNRAS.356..979M. doi:10.1111/j.1365-2966.2004.08514.x.

[22] Mackie, Glen (February 1, 2002). "To see the Universe in a Grain of Taranaki Sand" (http://astronomy.swin.edu.au/~gmackie/billions.html). Swinburne University. . Retrieved 2006-12-20.

[23] "Unveiling the Secret of a Virgo Dwarf Galaxy" (http://www.eso.org/outreach/press-rel/pr-2000/pr-12-00.html). ESO. 2000-05-03. . Retrieved 2007-01-03.

[24] "Hubble's Largest Galaxy Portrait Offers a New High-Definition View" (http://www.nasa.gov/mission_pages/hubble/science/hst_spiral_m10.html). NASA. 2006-02-28. . Retrieved 2007-01-03.

[25] Vergano, Dan (1 December 2010). "Universe holds billions more stars than previously thought" (http://www.usatoday.com/tech/science/space/2010-12-01-dwarf-stars_N.htm). *USA Today*. . Retrieved 14 December 2010.

[26] N. Mandolesi, P. Calzolari, S. Cortiglioni, F. Delpino, G. Sironi (1986). "Large-scale homogeneity of the Universe measured by the microwave background". *Letters to Nature* **319** (6056): 751–753. Bibcode 1986Natur.319..751M. doi:10.1038/319751a0.

[27] Hinshaw, Gary (November 29, 2006). "New Three Year Results on the Oldest Light in the Universe" (http://map.gsfc.nasa.gov/m_mm.html). NASA WMAP. . Retrieved 2006-08-10.

[28] Hinshaw, Gary (December 15, 2005). "Tests of the Big Bang: The CMB" (http://map.gsfc.nasa.gov/m_uni/uni_101bbtest3.html). NASA WMAP. . Retrieved 2007-01-09.

[29] Rindler (1977), p. 202.

[30] Hinshaw, Gary (February 10, 2006). "What is the Universe Made Of?" (http://map.gsfc.nasa.gov/m_uni/uni_101matter.html). NASA WMAP. . Retrieved 2007-01-04.

[31] "Five-Year Wilkinson Microwave Anisotropy Probe (WMAP) Observations: Data Processing, Sky Maps, and Basic Results" (http://lambda.gsfc.nasa.gov/product/map/dr3/pub_papers/fiveyear/basic_results/wmap5basic.pdf) (PDF). nasa.gov. . Retrieved 2008-03-06.

[32] Wright EL (2005). "Age of the Universe" (http://www.astro.ucla.edu/~wright/age.html). UCLA. . Retrieved 2007-01-08.
Krauss LM, Chaboyer B (3 January 2003). "Age Estimates of Globular Clusters in the Milky Way: Constraints on Cosmology". *Science* **299** (5603): 65–69. Bibcode 2003Sci...299...65K. doi:10.1126/science.1075631. PMID 12511641.

[33] Wright, Edward L. (September 12, 2004). "Big Bang Nucleosynthesis" (http://www.astro.ucla.edu/~wright/BBNS.html). UCLA. . Retrieved 2007-01-05.
M. Harwit, M. Spaans (2003). "Chemical Composition of the Early Universe". *The Astrophysical Journal* **589** (1): 53–57. arXiv:astro-ph/0302259. Bibcode 2003ApJ...589...53H. doi:10.1086/374415.
C. Kobulnicky, E. D. Skillman; Skillman (1997). "Chemical Composition of the Early Universe". *Bulletin of the American Astronomical Society* **29**: 1329. Bibcode 1997AAS...191.7603K.

[34] "Antimatter" (http://www.pparc.ac.uk/ps/bbs/bbs_antimatter.asp). Particle Physics and Astronomy Research Council. October 28, 2003. . Retrieved 2006-08-10.

[35] Landau and Lifshitz (1975), p. 361.

[36] WMAP Mission: Results – Age of the Universe (http://map.gsfc.nasa.gov/m_mm/mr_content.html). Map.gsfc.nasa.gov. Retrieved on 2011-11-28.

[37] Luminet, Jean-Pierre; Boudewijn F. Roukema (1999). "Topology of the Universe: Theory and Observations". *Proceedings of Cosmology School held at Cargese, Corsica, August 1998*. arXiv:astro-ph/9901364.
Luminet, Jean-Pierre; J. Weeks, A. Riazuelo, R. Lehoucq, J.-P. Uzan (2003). "Dodecahedral space topology as an explanation for weak wide-angle temperature correlations in the cosmic microwave background". *Nature* **425** (6958): 593–595. arXiv:astro-ph/0310253. Bibcode 2003Natur.425..593L. doi:10.1038/nature01944. PMID 14534579.

[38] Strobel, Nick (May 23, 2001). "The Composition of Stars" (http://www.astronomynotes.com/starprop/s7.htm). Astronomy Notes. . Retrieved 2007-01-04.
"Have physical constants changed with time?" (http://www.faqs.org/faqs/astronomy/faq/part4/section-4.html). Astrophysics (Astronomy Frequently Asked Questions). . Retrieved 2007-01-04.

[39] Stephen Hawking (1988). *A Brief History of Time*. Bantam Books. p. 125. ISBN 0-553-05340-X.

[40] Martin Rees (1999). *Just Six Numbers*. HarperCollins Publishers. ISBN 0-465-03672-4.

[41] Adams, F.C. (2008). "Stars in other universes: stellar structure with different fundamental constants". *Journal of Cosmology and Astroparticle Physics* **2008** (8): 010. arXiv:0807.3697. Bibcode 2008JCAP...08..010A. doi:10.1088/1475-7516/2008/08/010.

[42] Harnik, R.; Kribs, G.D. and Perez, G. (2006). "A universe without weak interactions". *Physical Review D* **74** (3): 035006. arXiv:hep-ph/0604027. Bibcode 2006PhRvD..74c5006H. doi:10.1103/PhysRevD.74.035006.

[43] Will Durant, *Our Oriental Heritage*:

> "Two systems of Hindu thought propound physical theories suggestively similar to those of Greece. Kanada, founder of the Vaisheshika philosophy, held that the world was composed of atoms as many in kind as the various elements. The Jains more nearly approximated to Democritus by teaching that all atoms were of the same kind, producing different effects by diverse modes of combinations. Kanada believed light and heat to be varieties of the same substance; Udayana taught that all heat comes from the sun; and Vachaspati, like Newton, interpreted light as composed of minute particles emitted by substances and striking the eye."

[44] Stcherbatsky, F. Th. (1930, 1962), *Buddhist Logic*, Volume 1, p. 19, Dover, New York:

> "The Buddhists denied the existence of substantial matter altogether. Movement consists for them of moments, it is a staccato movement, momentary flashes of a stream of energy... "Everything is evanescent",... says the Buddhist, because there is no stuff... Both systems [Sānkhya, and later Indian Buddhism] share in common a tendency to push the analysis of Existence up to its minutest, last elements which are imagined as absolute qualities, or things possessing only one unique quality. They are called "qualities" (guna-dharma) in both systems in the sense of absolute qualities, a kind of atomic, or intra-atomic, energies of which the empirical things are composed. Both systems, therefore, agree in denying the objective reality of the categories of Substance and Quality,... and of the relation of Inference uniting them. There is in Sānkhya philosophy no separate existence of qualities. What we call

quality is but a particular manifestation of a subtle entity. To every new unit of quality corresponds a subtle quantum of matter which is called guna "quality", but represents a subtle substantive entity. The same applies to early Buddhism where all qualities are substantive... or, more precisely, dynamic entities, although they are also called dharmas ('qualities')."

[45] Craig, William Lane (June 1979). "Whitrow and Popper on the Impossibility of an Infinite Past". *The British Journal for the Philosophy of Science* **30** (2): 165–170 (165–6). doi:10.1093/bjps/30.2.165.

[46] Boyer, C. *A History of Mathematics*. Wiley, p. 54.

[47] Neugebauer, Otto E. (1945). "The History of Ancient Astronomy Problems and Methods". *Journal of Near Eastern Studies* **4** (1): 1–38. doi:10.1086/370729. JSTOR 595168. "the Chaldaean Seleucus from Seleucia"

[48] Sarton, George (1955). "Chaldaean Astronomy of the Last Three Centuries B. C". *Journal of the American Oriental Society* **75** (3): 166–173 (169). doi:10.2307/595168. JSTOR 595168. "the heliocentrical astronomy invented by Aristarchos of Samos and still defended a century later by Seleucos the Babylonian"

[49] William P. D. Wightman (1951, 1953), *The Growth of Scientific Ideas*, Yale University Press p. 38, where Wightman calls him Seleukos the Chaldean.

[50] Lucio Russo, *Flussi e riflussi*, Feltrinelli, Milano, 2003, ISBN 88-07-10349-4.

[51] Bartel, p. 527

[52] Bartel, pp. 527–9

[53] Bartel, pp. 529–34

[54] Bartel, pp. 534–7

[55] Nasr, Seyyed H. (1st edition in 1964, 2nd edition in 1993). *An Introduction to Islamic Cosmological Doctrines* (2nd ed.). 1st edition by Harvard University Press, 2nd edition by State University of New York Press. pp. 135–6. ISBN 0791415155.

[56] Misner, Thorne and Wheeler (1973), p. 754.

[57] Misner, Thorne, and Wheeler (1973), p. 755–756.

[58] Misner, Thorne, and Wheeler (1973), p. 756.

[59] de Cheseaux JPL (1744). *Traité de la Comète*. Lausanne. pp. 223ff.. Reprinted as Appendix II in Dickson FP (1969). *The Bowl of Night: The Physical Universe and Scientific Thought*. Cambridge, MA: M.I.T. Press. ISBN 978-0262540032.

[60] Olbers HWM (1826). "Unknown title". *Bode's Jahrbuch* **111**.. Reprinted as Appendix I in Dickson FP (1969). *The Bowl of Night: The Physical Universe and Scientific Thought*. Cambridge, MA: M.I.T. Press. ISBN 978-0262540032.

[61] Jeans, J. H. (1902). "The Stability of a Spherical Nebula" (http://maeresearch.ucsd.edu/~cgibson/Documents2007/PapersAList copy/MiscellaneousPapers/Jeans1902.pdf) (PDF). *Philosophical Transactions Royal Society of London, Series A* **199** (312–320): 1–53. Bibcode 1902RSPTA.199....1J. doi:10.1098/rsta.1902.0012. JSTOR 90845. . Retrieved 17 March 2011.

[62] Rindler, p. 196; Misner, Thorne, and Wheeler (1973), p. 757.

[63] Misner, Thorne and Wheeler, p. 756.

[64] Einstein, A (1917). "Kosmologische Betrachtungen zur allgemeinen Relativitätstheorie". *Preussische Akademie der Wissenschaften, Sitzungsberichte*. 1917 (**part 1**): 142–152.

[65] Rindler (1977), pp. 226–229.

[66] Landau and Lifshitz (1975), pp. 358–359.

[67] Einstein, A (1931). "Zum kosmologischen Problem der allgemeinen Relativitätstheorie". *Sitzungsberichte der Preussischen Akademie der Wissenschaften, Physikalisch-mathematische Klasse* **1931**: 235–237.
Einstein A., de Sitter W. (1932). "On the relation between the expansion and the mean density of the universe". *Proceedings of the National Academy of Sciences* **18** (3): 213–214. Bibcode 1932PNAS...18..213E. doi:10.1073/pnas.18.3.213. PMC 1076193. PMID 16587663.

[68] Hubble Telescope news release (http://hubblesite.org/newscenter/archive/releases/2004/12/text/). Hubblesite.org (2004-02-20). Retrieved on 2011-11-28.

[69] "Mysterious force's long presence" (http://news.bbc.co.uk/1/hi/sci/tech/6156110.stm). *BBC News*. 2006-11-16. .

[70] Zel'dovich YB (1967). "Cosmological constant and elementary particles". *Zh. Eksp. & Teor. Fiz. Pis'ma* **6**: 883–884. English translation in *Sov. Phys. — JTEP Lett.*, **6**, pp. 316–317 (1967).

[71] Friedmann A. (1922). "Über die Krümmung des Raumes". *Zeitschrift für Physik* **10** (1): 377–386. Bibcode 1922ZPhy...10..377F. doi:10.1007/BF01332580.

[72] Munitz MK (1959). "One Universe or Many?". *Journal of the History of Ideas* **12** (2): 231–255. doi:10.2307/2707516. JSTOR 2707516.

[73] Misner, Thorne and Wheeler (1973), p.753.

[74] Linde A. (1986). "Eternal chaotic inflation". *Mod. Phys. Lett.* **A1** (2): 81–85. Bibcode 1986MPLA....1...81L. doi:10.1142/S0217732386000129.
Linde A. (1986). "Eternally existing self-reproducing chaotic inflationary universe" (http://www.stanford.edu/~alinde/Eternal86.pdf) (PDF). *Phys. Lett.* **B175** (4): 395–400. Bibcode 1986PhLB..175..395L. doi:10.1016/0370-2693(86)90611-8. . Retrieved 17 March 2011.

[75] Tegmark M. (2003). "Parallel universes. Not just a staple of science fiction, other universes are a direct implication of cosmological observations" (http://www.scientificamerican.com/article.cfm?id=parallel-universes). *Scientific American* **288** (5): 40–51. PMID 12701329. .

[76] Max Tegmark (2003). "Parallel Universes". *In "Science and Ultimate Reality: from Quantum to Cosmos", honoring John Wheeler's 90th birthday. J. D. Barrow, P.C.W. Davies, & C.L. Harper eds. Cambridge University Press (2003)*: 2131. arXiv:astro-ph/0302131. Bibcode 2003astro.ph..2131T.

[77] Shape of the Universe (http://map.gsfc.nasa.gov/universe/uni_shape.html), WMAP website at NASA.

[78] Luminet, Jean-Pierre; Jeff Weeks, Alain Riazuelo, Roland Lehoucq, Jean-Phillipe Uzan (2003-10-09). "Dodecahedral space topology as an explanation for weak wide-angle temperature correlations in the cosmic microwave background". *Nature* **425** (6958): 593–5. arXiv:astro-ph/0310253. Bibcode 2003Natur.425..593L. doi:10.1038/nature01944. PMID 14534579.

[79] Roukema, Boudewijn; Zbigniew Buliński, Agnieszka Szaniewska, Nicolas E. Gaudin (2008). "A test of the Poincare dodecahedral space topology hypothesis with the WMAP CMB data". *Astronomy and Astrophysics* **482** (3): 747. arXiv:0801.0006. Bibcode 2008A&A...482..747L. doi:10.1051/0004-6361:20078777.

[80] Aurich, Ralf; Lustig, S., Steiner, F., Then, H. (2004). "Hyperbolic Universes with a Horned Topology and the CMB Anisotropy". *Classical and Quantum Gravity* **21** (21): 4901–4926. arXiv:astro-ph/0403597. Bibcode 2004CQGra..21.4901A. doi:10.1088/0264-9381/21/21/010.

## Bibliography

- Bartel (1987). "The Heliocentric System in Greek, Persian and Hindu Astronomy". *Annals of the New York Academy of Sciences* **500** (1): 525–545. Bibcode 1987NYASA.500..525V. doi:10.1111/j.1749-6632.1987.tb37224.x.
- Landau, Lev, Lifshitz, E.M. (1975). *The Classical Theory of Fields (Course of Theoretical Physics, Vol. 2)* (revised 4th English ed.). New York: Pergamon Press. pp. 358–397. ISBN 9780080181769.
- Liddell, H. G. and Scott, R. *A Greek-English Lexicon*, Oxford University Press, ISBN 0-19-864214-8
- Misner, C.W., Thorne, Kip, Wheeler, J.A. (1973). *Gravitation*. San Francisco: W. H. Freeman. pp. 703–816. ISBN 978-0-7167-0344-0.
- Rindler, W. (1977). *Essential Relativity: Special, General, and Cosmological*. New York: Springer Verlag. pp. 193–244. ISBN 0-387-10090-3.

## Further reading

- Weinberg, S. (1993). *The First Three Minutes: A Modern View of the Origin of the Universe* (2nd updated ed.). New York: Basic Books. ISBN 978-0465024377. OCLC 28746057. For lay readers.
- Nussbaumer, Harry; Bieri, Lydia; Sandage, Allan (2009). *Discovering the Expanding Universe* (http://books.google.com/books?id=RaNOJkQ4l14C). Cambridge University Press. ISBN 978-0-521-51484-2.

## External links

- Is there a hole in the universe? (http://www.howstuffworks.com/hole-in-universe.htm) at HowStuffWorks
- *Stephen Hawking's Universe* (http://www.pbs.org/wnet/hawking/html/home.html) – Why is the universe the way it is?
- Cosmology FAQ (http://www.astro.ucla.edu/~wright/cosmology_faq.html)
- Cosmos – An "illustrated dimensional journey from microcosmos to macrocosmos" (http://www.shekpvar.net/~dna/Publications/Cosmos/cosmos.html)
- Illustration comparing the sizes of the planets, the sun, and other stars (http://www.co-intelligence.org/newsletter/comparisons.html)
- Logarithmic Maps of the Universe (http://www.astro.princeton.edu/~mjuric/universe/)
- My So-Called Universe (http://web.archive.org/web/20101225211703/http://www.slate.com/id/2087206/nav/navoa/) – Arguments for and against an infinite and parallel universes
- The Dark Side and the Bright Side of the Universe (http://cosmology.lbl.gov/talks/Ho_07.pdf) Princeton University, Shirley Ho
- Richard Powell: *An Atlas of the Universe* (http://www.atlasoftheuniverse.com/) – Images at various scales, with explanations
- Multiple Big Bangs (http://www.npr.org/templates/story/story.php?storyId=1142346)

- Universe – Space Information Centre (http://www.exploreuniverse.com/ic/)
- Exploring the Universe (http://www.nasa.gov/topics/universe/index.html) at Nasa.gov

### Videos

- The Known Universe (http://www.youtube.com/embed/17jymDn0W6U) created by the American Museum of Natural History
- Understand The Size Of The Universe (http://www.youtube.com/embed/0fKBhvDjuy0) – by Powers of Ten

# Polytheism

**Polytheism** is the belief of multiple deities also usually assembled into a pantheon of gods and goddesses, along with their own mythologies and rituals.

Egyptians Gods

Polytheism was the typical form of religion during the Bronze Age and Iron Age, up to the Axial Age and the gradual development of monotheism or pantheism, and atheism. It is well documented in historical religions of Classical Antiquity, especially Greek polytheism and Roman polytheism, and after the decline of classical polytheism in tribal religions such as Germanic polytheism or Slavic polytheism. It persists into the modern period in traditions such as Hinduism, Buddhism, Shintoism, Chinese folk religion, etc., and it has been revived in currents of Neopaganism in the post-Christian West.

Polytheism is a type of theism. Within theism, it contrasts with monotheism, the belief in a singular God. Polytheists do not always worship all the gods equally, but can be Henotheists, specializing in the worship of one particular deity. Other polytheists can be Kathenotheists, worshipping different deities at different times.

## Gods and divinity

The deities of polytheistic religions are agents in mythology, where they are portrayed as complex personages of greater or lesser status, with individual skills, needs, desires and histories. These gods are often seen as similar to humans (anthropomorphic) in their personality traits, but with additional individual powers, abilities, knowledge or perceptions. Polytheism cannot be cleanly separated from the animist beliefs prevalent in most folk religions. The gods of polytheism are in many cases the highest order of a continuum of supernatural beings or spirits, which may include ancestors, demons, wights and others. In some cases these spirits are divided into celestial or chthonic classes, and belief in the existence of all these beings does not imply that all are worshipped.

## Types of deities

Further information: List of deities

Types of deities often found in polytheism

- Creator deity
- Culture hero
- Death deity (chthonic)
- Life-death-rebirth deity
- Love goddess
- Mother goddess
- Political deity (such as a king or emperor)
- Sky deity (celestial)
- Solar deity
- Trickster deity
- Water deity
- Gods of music, arts, science, farming or other endeavors.

## Mythology and religion

In the Classical era, Sallustius (4th century CE) categorised mythology into five types:

1. Theological
2. Physical
3. Psychological
4. Material
5. Mixed

The theological are those myths which use no bodily form but contemplate the very essence of the gods: e.g., Cronus swallowing his children. Since divinity is intellectual, and all intellect returns into itself, this myth expresses in allegory the essence of divinity.

Myths may be regarded physically when they express the activities of gods in the world.

The psychological way is to regard (myths as allegories of) the activities of the soul itself and or the soul's acts of thought.

The material is to regard material objects to actually be gods, for example: to call the earth Gaia, ocean Okeanos, or heat Typhon.

The mixed kind of myth may be seen in many instances: for example they say that in a banquet of the gods, Eris threw down a golden apple; the goddesses contended for it, and were sent by Zeus to Paris to be judged. Paris saw Aphrodite to be beautiful and gave her the apple. Here the banquet signifies the hypercosmic powers of the gods; that is why they are all together. The golden apple is the world, which being formed out of opposites, is naturally said to be 'thrown by Eris ' (or Discord). The different gods bestow different gifts upon the world, and are thus said to 'contend for the apple'. And the soul which lives according to sense - for that is what Paris is - not seeing the other powers in the world but only beauty, declares that the apple belongs to Aphrodite.

## Historical polytheism

Some well-known historical polytheistic pantheons include the Sumerian gods and the Egyptian gods, and the classical-attested pantheon which includes the ancient Greek religion and Roman religion. Post-classical polytheistic religions include Norse Æsir and Vanir, the Yoruba Orisha, the Aztec gods, and many others. Today, most historical polytheistic religions are pejoratively referred to as "mythology", though the stories cultures tell about their gods should be distinguished from their worship or religious practice. For instance deities portrayed in conflict in mythology would still be worshipped sometimes in the same temple side by side, illustrating the distinction in the devotees mind between the myth and the reality. It is speculated that there was once a Proto-Indo-European religion, from which the religions of the various Indo-European peoples derive, and that this religion was an essentially naturalist numenistic religion. An example of a religious notion from this shared past is the concept of **dyēus*, which is attested in several distinct religious systems.

In many civilizations, pantheons tended to grow over time. Deities first worshipped as the patrons of cities or places came to be collected together as empires extended over larger territories. Conquests could lead to the subordination of the elder culture's pantheon to a newer one, as in the Greek Titanomachia, and possibly also the case of the Æsir and Vanir in the Norse mythos. Cultural exchange could lead to "the same" deity being renowned in two places under different names, as with the Greeks, Etruscans, and Romans, and also to the introduction of elements of a "foreign" religion into a local cult, as with Egyptian Osiris worship brought to ancient Greece.

Most ancient belief systems held that gods influenced human lives. However, the Greek philosopher Epicurus held that the gods were living, incorruptible, blissful beings who did not trouble themselves with the affairs of mortals, but who could be perceived by the mind, especially during sleep. Epicurus believed that these gods were material, human-like, and that they inhabited the empty spaces between worlds.

Hellenistic religion may still be regarded as polytheistic, but with strong monistic components, and monotheism finally emerges out of Hellenistic traditions in Late Antiquity in the form of Neoplatonism and Christian theology.

Bronze Age to Classical Antiquity

- Religions of the Ancient Near East
  - Ancient Egyptian religion
  - Ancient Semitic religion
- Historical Vedic religion
- Ancient Greek religion
- Ancient Roman religion
- Celtic polytheism

Late Antiquity to High Middle Ages

- Germanic paganism
- Slavic paganism
- Baltic paganism
- Finnish paganism

### Ancient Greece

The classical scheme in Ancient Greece of the Twelve Olympians (the Canonical Twelve of art and poetry) were:[1][2] Zeus, Hera, Poseidon, Athena, Ares, Demeter, Apollo, Artemis, Hephaestus, Aphrodite, Hermes and Dionysus; Hestia stepped down when Dionysus was offered a seat on Mount Olympus and Hades[3] was often excluded because he dwelt in the underworld. All of the gods had a power. There was, however, a great deal of fluidity as to whom was counted among their number in antiquity.[4] Different cities often worshipped the same deities, sometimes with epithets that distinguished them and specified their local nature.

The Hellenic Polytheism extended beyond mainland Greece, to the islands and coasts of Ionia in Asia Minor, to Magna Graecia (Sicily and southern Italy), and to scattered Greek colonies in the Western Mediterranean, such as Massalia (Marseille). Greek religion tempered Etruscan cult and belief to form much of the later Roman religion.

## Folk religion

Further information: Saint, Angel, Folk Catholicism, and Shamanism

The animistic nature of folk beliefs is an anthropological cultural universal. The belief in ghosts and spirits animating the natural world and the practice of ancestor worship is universally present in the world's cultures and re-emerges in monotheistic or materialistic societies as "superstition", belief in demons, tutelary saints, fairies or extraterrestrials.

The presence of a full polytheistic religion, complete with a ritual cult conducted by a priestly caste, requires a higher level of organization and is not present in every culture. Explicit polytheism in contemporary folk religion is found in African traditional religion as well as African diasporic religions. In Eurasia, the Kalash are one of very few instances of surviving polytheism. Also, a large number of polytheist folk traditions are subsumed in contemporary Hinduism, although Hinduism is doctrinally dominated by monist or monotheist theology (Bhakti, Advaita). Historical Vedic polytheist ritualism survives as a minor current in Hinduism, known as Shrauta. More widespread is folk Hinduism, with rituals dedicated to various local or regional deities.

## Contemporary world religions

### Hinduism

Further information: Hindu views on monotheism

Religious Hinduism is a broad category which encompasses both monotheistic and polytheistic tendencies and variations on or mixes of both structures, Hinduism is accused of polytheism but Hinduism cannot be considered polytheistic as all great Hindu religious leaders have repeatedly stressed that god is one and his forms are many,the ways to communicate with him are many and focussing or concentrating on the icon is one of those ways. For a Hindu the human language itself is a symbolic representation of the divine.

Hindus venerate God in the form of the Murti, an icon. the Puja (worship) of the Murti is like a way to communicate with the formless, abstract divinity (Brahman in Hinduism) which creates, sustains and dissolves creation.

Hindu philosophers and theologians also argue for a transcendent metaphysical structure with a single divine essence. This divine essence is usually referred to as Brahman or Atman, but the understanding of the nature of this absolute divine essence is the line which defines many Hindu philosophical traditions such as Vedanta. Hindus believe that the decision of whose path to god is correct will be decided by god only, so everything must be left to and surrendered unto god and no one should take the matters into his own hands or use violence.

Many Hindus believe in different deities emanating from Brahman, and the majority continues to worship a deity as a matter of personal belief or tradition as a representation of this supreme being, as a representation of the 'One God'.

In the Smartha denomination of Hinduism, the philosophy of Advaita expounded by Shankara allows veneration of numerous deities with the understanding that all of them are but manifestations of one impersonal divine power, Brahman. Therefore, according to various schools of Vedanta including Shankara, which is the most influential and important Hindu theological tradition, there are a great number of deities in Hinduism, such as Vishnu, Shiva, Ganesha, Hanuman, Lakshmi, and Kali, but they are essentially different forms of the same "Being". However, many Vedantic philosophers also argue that all individuals were united by the same impersonal, divine power in the form of the Atman.

Many other Hindus, however, view polytheism as far preferable to monotheism. Ram Swarup, for example, points to the Vedas as being specifically polytheistic,[5] and states that, "only some form of polytheism alone can do justice to this variety and richness."[6] Sita Ram Goel, another 20th century Hindu historian, wrote:

> "I had an occasion to read the typescript of a book [Ram Swarup] had finished writing in 1973. It was a profound study of Monotheism, the central dogma of both Islam and Christianity, as well as a powerful presentation of what the monotheists denounce as Hindu Polytheism. I had never read anything like it. It was a revelation to me that Monotheism was not a religious concept but an imperialist idea. I must confess that I myself had been inclined towards Monotheism till this time. I had never thought that a multiplicity of Gods was the natural and spontaneous expression of an evolved consciousness."[7]

### Buddhism and Shinto

Further information: God in Buddhism, Devas vs. Gods, and Nontheism in Buddhism

In Buddhism, there are higher beings commonly designed (or designated) as gods, Devas; however, Buddhism, at its core (the original Pali canon), does not teach the notion of praying nor worship to the Devas or any god(s).

However, there are many gods in Buddhism who end being worshipped in practice. Worshipping images of Buddha is common, Buddhism has also integrated with the Chinese Folk Religion and many gods from that religion are considered Bodhisattvas which means incarnations of the Buddha.[8] [9]

Devas, in general, are beings who have had more positive karma in their past lives than humans. Their lifespan eventually ends. When their lives end, they will be reborn as devas or as other beings. When they accumulate negative karma, they are reborn as either human or any of the other lower beings. Humans and other beings could also be reborn as a deva in their next rebirth, if they accumulate enough positive karma; however, it is not recommended.

Buddhism flourished in different countries, and some of those countries have polytheistic folk religions. Buddhism syncretizes easily with other religions. Thus, Buddhism has mixed with the folk religions and emerged in polytheistic variants as well as nontheistic variants. For example, in Japan, Buddhism, mixed with Shinto, which worships deities called kami, created a tradition which prays to the deities of Shinto as a form of Buddha. Thus, there may be elements of worship of gods in some forms of later Buddhism.

The concepts of Adi-Buddha and Dharmakaya are the closest to monotheism any form of Buddhism comes. All famous sages and Bodhisattvas being considered as reflections of it. Adi-Buddha is not said to be the creator, but the originator of all things, being a deity in an Emanationist sense.

## Neopaganism

Further information: Theology, Pantheon (gods), Euhemerism, Interpretatio graeca, Demigod, and Apotheosis

*Hard Polytheists* believe that gods are distinct, **separate** real divine beings not psychological archetypes or personifications of natural forces. Hard polytheists reject the idea that "all gods are one God"

This is contrasted with *Soft Polytheism*, which holds that Gods may be aspects of only one God, psychological archetypes or personifications of natural forces.

It is a misconception that *Hard Polytheists* consider the gods of all cultures as being equally real; that is a theological position more correctly called integrational polytheism or omnitheism.

*Soft Polytheism* is prevalent in New Age and syncretic currents of Neopaganism, as are psychological interpretations of deities as archetypes of the human psyche. English occultist Dion Fortune was a major populiser of soft polytheism. In her novel, *The Sea Priestess*, she wrote, "All gods are one god, and all goddesses are one goddess, and there is one initiator."[10] This phrase is very popular among some Neopagans (notably, Wiccans) and incorrectly often believed to be just a recent work of fiction. However, Fortune indeed quoted from an ancient source, the Latin novel The Golden Ass of Apuleius. Fortune's soft polytheist compromise between monotheism and polytheism has been described as "pantheism" (Greek: πάν *pan* 'all' and θεός *theos* 'god'). However, "Pantheism" has a longer history of usage to refer to a view of an all-encompassing immanent divine.

Neopaganism often blends polytheism with pantheism or panentheism.

### Wicca

Wicca is a pantheistic, duotheistic, and a polytheistic faith. It sees the universe as being comprised by a divine Godhead (known by various names), but who is subdivided into the opposing polarities of The God and The Goddess. Each of these deities can be further divided into many different polytheistic deities, which are aspects of The God and The Goddess. Wicca is tolerant in the understanding of divinity, but emphasises a balance and equality between male and female deities, whereas other polytheistic faiths have often placed male deities at the top of the hierarchy.

### Reconstructionism

Reconstructionists are neopagans which apply scholarly disciplines such as History, Etymology, Archaeology, Linguistics and others to a traditional religion that has been destroyed such as Norse Paganism, Greek Paganism, Celtic Paganism and others. After researching his or her path a reconstructionist or "recon" for short will apply the customs, morals and worldview to the modern day. In general they are considered "Hard Polytheists" and value historical research over personal insight.

## See also

- Apotheosis
- Henotheism
- Judgement of Paris
- Monotheism
- Mythology
- Paganism
- Pantheism
- Polytheistic reconstructionism
- Shirk (polytheism)
- Theism

## References

[1] "Greek mythology". *Encyclopedia Americana*. **13**. 1993. p. 431.

[2] "Dodekatheon [Twelve Olympians]" (in Greek). *Papyrus Larousse Britannica*. 2007.

[3] George Edward Rines, ed (1919). "Encyclopedia Americana Vol. 13" (http://books.google.com/books?id=PWYMAAAAYAAJ&pg=PA408). *Encyclopedia Americana Vol. 13*. Americana Corp. pp. 408–411. .

[4] Stoll, Heinrich Wilhelm (R.B. Paul trans.) (1852). *Handbook of the religion and mythology of the Greeks*. Francis and John Rivington. p. 8. "The limitation [of the number of Olympians] to twelve seems to have been a comparatively modern idea"

[5] Goel, Sita Ram (1987). *Defence of Hindu Society* (http://voi.org/books/hindusoc/ch5.htm). New Delhi, India: Voice of India. . ""In the Vedic approach, there is no single God. This is bad enough. But the Hindus do not have even a supreme God, a fuhrer-God who presides over a multiplicity of Gods." -- Ram Swarup"

[6] Goel, Sita Ram (1987). *Defence of Hindu Society* (http://voi.org/books/hindusoc/ch5.htm). New Delhi, India: Voice of India. .

[7] Goel, Sita Ram (1982). *How I became a Hindu*. New Delhi, India: Voice of India. pp. 92.

[8] http://www.oldpaths.com/archive/grubb/john/mark/1953/buddhism.html

[9] http://www.harunyahya.com/buddhism02.php

[10] Fortune, Dion; Knight, Gareth (30 June 2003). *The Sea Priestess* (http://books.google.com/books?id=uZl4DKIB5WoC&pg=PP1). Weiser. p. 169. ISBN 9781578632909. . "All gods are one god, and all goddesses are one goddess, and there is one initiator."

## Further reading


- Assmann, Jan, 'Monotheism and Polytheism' in: Sarah Iles Johnston (ed.), *Religions of the Ancient World: A Guide*, Harvard University Press (2004), ISBN 0-674-01517-7, pp. 17–31.
- Burkert, Walter, *Greek Religion: Archaic and Classical*, Blackwell (1985), ISBN 0-631-15624-0.
- Greer, John Michael; *A World Full of Gods: An Inquiry Into Polytheism*, ADF Publishing (2005), ISBN 0-9765681-0-1
- Iles Johnston, Sarah; *Ancient Religions*, Belknap Press (September 15, 2007), ISBN 0-674-02548-2
- Marbaniang, Domenic *Epistemics of Divine Reality* Google Books (http://books.google.co.in/books?id=fL-dAqxShiwC&pg=PP1&dq=Epistemics+of+Divine+Reality) (See Chapter 3 Empirical Epistemics of Divine Reality for philosophical analysis of polytheism)
- Paper, Jordan; *The Deities are Many: A Polytheistic Theology*, State University of New York Press (March 3, 2005), ISBN 978-0-7914-6387-1
- Penchansky, David, *Twilight of the Gods: Polytheism in the Hebrew Bible* (2005), ISBN 0-664-22885-2.


## External links

- The Association of Polytheist Traditions (http://www.manygods.org.uk/) - APT, a UK-based community of Polytheists.
- International Year Of Polytheism (http://www.monochrom.at/polytheism) Philosophical project promoting polytheism by group monochrom
- Integrational Polytheism (http://www.polytheism.org.uk/wordpress/)

# Paganism

***Paganism*** (from Latin *paganus*, meaning "country dweller", "rustic"[1]) is a blanket term, typically used to refer to non-Abrahamic, indigenous polytheistic religious traditions.

Venus of Arles, late 1st century BCE.

It is primarily used in a historical context, referring to Greco-Roman polytheism as well as the polytheistic traditions of Europe and North Africa before Christianization. In a wider sense, extended to contemporary religions, it includes most of the Eastern religions and the indigenous traditions of the Americas, Central Asia, Australia and Africa; as well as non-Abrahamic folk religion in general. More narrow definitions will not include any of the world religions and restrict the term to local or rural currents not organized as civil religions. Characteristic of Pagan traditions is the absence of proselytism and the presence of a living mythology, which explains religious practice.

Isis holding a sistrum and an oinochoe. (Roman artwork from the Hadrian period (117–138 CE).)

In the Christian perspective the term has been used historically to encompass all non–Abrahamic religions.[2] [3] The term *pagan* is a Christian adaptation of the "gentile" of Judaism, and as such has an inherent Abrahamic bias, and pejorative connotations among monotheists,[4] comparable to heathen and infidel also known as kafir (كافر) and mushrik in Islam. Peter Brown observes:

> The adoption of *paganus* by Latin Christians as an all-embracing, pejorative term for polytheists represents an unforeseen and singularly long-lasting victory, within a religious group, of a word of Latin slang originally devoid of religious meaning. The evolution occurred only in the Latin west, and in connection with the Latin church. Elsewhere, "Hellene " or "gentile" (*ethnikos*) remained the word for "pagan"; and *paganos* continued as a purely secular term, with overtones of the inferior and the commonplace.[5]

For these reasons, ethnologists avoid the term "paganism," with its uncertain and varied meanings, in referring to traditional or historic faiths, preferring more precise categories such as polytheism, shamanism, pantheism, or animism.

In the late 20th century, "Paganism", *or "Neopaganism",* became widely used in reference to adherents of various New Religious Movements including Wicca.[6] As such, various modern scholars have begun to apply the term to three groups of separate faiths: Historical Polytheism (such as Celtic polytheism, Norse Paganism, and Hellenic Polytheistic Reconstructionism also called Hellenismos), Folk/ethnic/Indigenous religions (such as Chinese folk religion and African traditional religion), and Neopaganism (such as Wicca and Germanic Neopaganism).

## Etymology

### Pagan

Reconstruction of the Parthenon, on the Acropolis of Athens, Greece.

The term *pagan* is from the Latin *paganus*, an adjective originally meaning "rural", "rustic" or "of the country." As a noun, *paganus* was used to mean "country dweller, villager."[7] The semantic development of post-classical Latin *paganus* in the sense "non-Christian, heathen" is unclear. The dating of this sense is controversial, but the 4th century seems most plausible. An earlier example has been suggested in Tertullian De Corona Militis xi, *"Apud hunc [sc. Christum] tam miles est paganus fidelis quam paganus est miles infidelis,"* but here the word *paganus* may be interpreted in the sense "civilian" rather than "heathen". There are three main explanations of the development:

The Parthenon in Nashville's Centennial Park is a full-scale copy of the original Greek Parthenon.

- (i) The older sense of classical Latin *pāgānus* is "of the country, rustic" (also as noun). It has been argued that the transferred use reflects the fact that the ancient idolatry lingered on in the rural villages and hamlets after Christianity had been accepted in the towns and cities of the Roman Empire;[8] cf. Orosius Histories 1. Prol. *"Ex locorum agrestium compitis et pagis pagani vocantur."* From its earliest beginnings, Christianity spread much more quickly in major urban areas (like Antioch, Alexandria, Carthage, Corinth, Rome) than in the countryside (in fact, the early church was almost entirely urban ), and soon the word for "country dweller" became synonymous with someone who was "not a Christian," giving rise to the modern meaning of "pagan." This may, in part, have had to do with the closeness to nature of rural people, who may have been more resistant to the new ideas of Christianity than those who lived in major urban centers and were cut off from the cycles of nature and the forms of spirituality associated with them. However, it may have also resulted from early Christian missionaries focusing their efforts within major population centers (e.g., St. Paul), rather than throughout an expansive, yet sparsely populated, countryside (hence, the Latin term suggesting "uneducated country folk") until a bit later on.
- (ii) The more common meaning of classical Latin *pāgānus* is "civilian, non-militant" (adjective and noun). Christians called themselves *mīlitēs*, "enrolled soldiers" of Christ, members of his militant church, and applied to non-Christians the term applied by soldiers to all who were "not enrolled in the army".
- (iii) The sense "heathen" arose from an interpretation of *paganus* as denoting a person who was outside a particular group or community, hence "not of the city" or "rural"; cf. Orosius Histories 1. Prol. *"ui alieni a civitate dei..pagani vocantur."* See C. Mohrmann, Vigiliae Christianae 6 (1952) 9ff.

—Oxford English Dictionary, (online) 2nd Edition (1989) [9]

The post-classical Latin *paganismus* gave rise to both *paganism* and to its synonym *paynimry*.[10] Paynimry may be used of paganism, its practises, and pagans,[11] as well as for the domain or realm of pagans.[12]

"Peasant" is a cognate, via Old French *paisent*.[13]

In their origins, these usages derived from *pagus*, "province, countryside", cognate to Greek πάγος "rocky hill", and, even earlier, "something stuck in the ground", as a landmark: the Proto-Indo-European root **pag-* means "fixed" and is also the source of the words *page*, *pale* (stake), and *pole*, as well as *pact* and *peace*.

While *pagan* is attested in English from the 14th century, there is no evidence that the term *paganism* was in use in English before the 17th century. The *OED* instances Edward Gibbon's *Decline and Fall of the Roman Empire* (1776): "The divisions of Christianity suspended the ruin of Paganism." The term was not a neologism, however, as *paganismus* was already used by Augustine.[14]

Less than twenty years after the last vestiges of Paganism were crushed with great severity by the emperor Theodosius I[15] Rome was seized by Alaric in 410. This led to murmuring that the gods of Paganism had taken greater care of the city than that of the Christian God, inspiring St Augustine to write *The City of God*, alternative title "*De Civitate Dei contra Paganos*: The City of God against the Pagans", in which he claimed that whilst the great 'city of Man' had fallen, Christians were ultimately citizens of the 'city of God.'[16]

### Heathen

*Heathen* is from Old English *hæðen* "not Christian or Jewish" (c.f. Old Norse *heiðinn*). Historically, the term was probably influenced by Gothic *haiþi* "dwelling on the heath", appearing as *haiþno* in Ulfilas' bible as "gentile woman" (translating the "Hellene" in Mark 7:26). This translation was probably influenced by Latin *paganus*, "country dweller", or it was chosen because of its similarity to the Greek ἐθνικός *ethnikos*, "gentile". It has even been suggested that Gothic *haiþi* is not related to "heath" at all, but rather a loan from Armenian *hethanos*, itself loaned from Greek ἔθνος *ethnos*.

## Terminology

Further information: Idolatry, Polytheism, and Ethnic religion

Both "pagan" and "heathen" have historically been used as a pejorative by adherents of monotheistic religions (such as Judaism, Christianity and Islam) to indicate a disbeliever in their religion; although in modern times it is not always used as a pejorative. "Paganism" frequently refers to the religions of classical antiquity, most notably Greek mythology or Roman religion; and can be used neutrally or admiringly by those who refer to those complexes of belief. However, until the rise of Romanticism and the general acceptance of freedom of religion in Western civilization, "paganism" was almost always used disparagingly of heterodox beliefs falling outside the established political framework of the Christian Church.

"Pagan" came to be equated with a Christianized sense of "epicurean" to signify a person who is sensual, materialistic, self-indulgent, unconcerned with the future and uninterested in sophisticated religion. The word was usually used in this worldly and stereotypical sense, particularly among those who were drawing attention to what they perceived as being the limitations of Paganism. Thus G. K. Chesterton wrote: "The Pagan set out, with admirable sense, to enjoy himself. By the end of his civilization he had discovered that a man cannot enjoy himself and continue to enjoy anything else." In sharp contrast, Swinburne the poet would comment on this same theme: "Thou hast conquered, O pale Galilean; the world has grown grey from thy breath; We have drunken of things Lethean, and fed on the fullness of death."[17]

Hypatia of Alexandria, philosopher, mathematician and astronomer, a pagan killed by a Christian mob in March 415 CE.

Christianity itself has been perceived at times as a form of polytheism by followers of the other Abrahamic religions[18] [19] because of the Christian doctrine of the Trinity (which at first glance appears indistinguishable from Tritheism,[20] though this is variously condemned as heresy or apostasy by the main Christian denominations) or the celebration of Pagan feast days[21] and other practices – through a process described as "baptizing"[22] or "christianization". Even between Christians there have been similar charges of idolatry levelled, especially by Protestants,[23] [24] towards the Roman Catholic and Orthodox Churches for their veneration of the saints and images.

## Historical Paganism

Further information: Prehistoric religion and Polytheism

### Bronze Age to Early Iron Age

- Religions of the Ancient Near East
  - Ancient Egyptian religion
  - Ancient Semitic religion
  - Mesopotamian religion
- reconstructed Proto-Indo-European religion
- Proto-Indo-Iranian religion
  - historical Vedic religion

### Classical Antiquity

Ludwig Feuerbach (1833) defines "Paganism" (*Heidentum*) in the context of classical antiquity as "the unity of religion and politics, of spirit and nature, of god and man",[25] qualified by the observation that "man" in the Pagan view is always defined by ethnicity, i.e. Greek, Roman, Egyptian, Jew, etc., so that each Pagan tradition is also a national tradition. Feuerbach goes on to postulate that the emergence of monotheism and thus the end of the Pagan period was a development which naturally grew out of Hellenistic philosophy due to the contradiction inherent in the ethnic nature of Pagan tradition and the universality of human spirituality (*Geist*), finally resulting in the emergence of a religion with a universalist scope in the form of Christianity.[26]

### Late Antiquity

Further information: Origins of Christianity, Decline of Hellenistic paganism, Magic in the Greco-Roman world, and Hellenic philosophy and Christianity

The developments of Late Antiquity in the religious thought in the far-flung Roman Empire needs to be addressed separately, as this is the context in which Early Christianity itself developed as one of several monotheistic cults, and it was in this period that the concept of "pagan" developed in the first place. Christianity as it emerged out of Second Temple Judaism (or Hellenistic Judaism) stood in competition with other religions advocating "pagan monotheism", including Neoplatonism, Mithraism, Gnosticism, Manichaeanism, and the cult of Dionysus.[27]

Dionysus in particular exhibits significant parallels with Christ, so that numerous scholars have concluded that the recasting of Jesus the wandering rabbi into the image of Christ the Logos, the divine saviour, reflects the cult of Dionysus directly. They point to the symbolism of wine and the importance it held in the mythology surrounding both Dionysus and Jesus Christ;[28] [29] Wick argues that the use of wine symbolism in the Gospel of John, including the story of the Marriage at Cana at which Jesus turns water into wine, was intended to show Jesus as superior to Dionysus.[30] The scene in *The Bacchae* wherein Dionysus appears before King Pentheus on charges of claiming divinity is compared to the New Testament scene of Jesus being interrogated by Pontius Pilate.[30] [31] [32]

For these reasons, it is difficult if not impossible to draw a clear line between "Christianity" and "Paganism" for the period of the 3rd to 4th centuries when Christianity was in its formative phase. Only with the emergence of Orthodox Christianity as reflected in the Apostle's Creed and the final decline of Hellenistic paganism by the 6th century does "Paganism" become a concept clearly distinct from Christianity.

## Ethnic religions of pre-Christian Europe

Further information: Christianization

(as opposed to Abrahamic religion)

- Celtic polytheism
- Germanic paganism
- Slavic paganism
- Baltic paganism
- Finnic paganism
- Vainakh paganism

# Pagan survivals in folklore

Further information: Folklore

In addition, folklore that is not any longer perceived as holding any religious significance can in some instances be traced to pre-Christian or pre-Islamic origins. In Europe, this is particularly the case with the various customs of Carnival or Fasnacht and the Yule traditions surrounding Santa Claus/Sinterklaas. By contrast, the Christmas tree in spite of frequent association with Thor's Oak cannot be shown to be an innovation predating the Early Modern period.

# Early Modern period

Interest in pagan traditions was revived in the Renaissance, at first in Renaissance magic as a revival of Greco-Roman magic. In the 17th century, description of paganism turned from the theological aspect to the ethnological, and a religion began to be understood as part of the ethnic identity of a people, and the study of the religions of "primitive" peoples triggered questions as to the ultimate historical origin of religion. Thus, Nicolas Fabri de Peiresc saw the pagan religions of Africa of his day as relicts that were in principle capable of shedding light on the historical Paganism of Classical Antiquity.[33]

## Romanticism

Paganism re-surfaces as a topic of fascination in 18th to 19th century Romanticism, in particular in the context of the literary Celtic and Viking revivals, which portrayed historical Celtic and Germanic polytheists as noble savages.

The 19th century also saw much scholarly interest in the reconstruction of pagan mythology from folklore or fairy tales. This was notably attempted by the Brothers Grimm, especially Jacob Grimm in his *Teutonic Mythology*, and Elias Lönnrot with the compilation of the *Kalevala*. The work of the Brothers Grimm influenced other collectors, both inspiring them to collect tales and leading them to similarly believe that the fairy tales of a country were particularly representative of it, to the neglect of cross-cultural influence. Among those influenced were the Russian Alexander Afanasyev, the Norwegians Peter Christen Asbjørnsen and Jørgen Moe, and the Englishman Joseph Jacobs.[34]

Romanticist interest in non-classical antiquity coincided with the rise of Romantic nationalism and the rise of the nation state in the context of the 1848 revolutions, leading to the creation of *national epics* and national myths for the various newly formed states. Pagan or folkloristic topics were also common in the Musical nationalism of the period.

## Contemporary Paganism

Contemporary Paganism, or Neopaganism, includes reconstructed religions such as Hellenic polytheism, Slavic neopaganism (i.e. Slavianstvo, including Rodnovery), Celtic Reconstructionist Paganism, or Germanic religious reconstructionism, as well as modern eclectic traditions such as Discordianism, Wicca and its many offshoots.

Romuvan priestess conducting a ceremony in Lithuania. Romuva is the Neopagan revival of the ethnic religion of the Lithuanians and the Balts.

Many of the "revivals", Wicca and Neo-druidism in particular, have their roots in 19th century Romanticism and retain noticeable elements of occultism or theosophy that were current then, setting them apart from historical rural (*paganus*) folk religion.

Pagan handfasting ceremony at Avebury (Beltane 2005).

Neopaganism in the United States accounts for roughly a third of all contemporary Pagans worldwide, and for some 0.2% of US population, figuring as the sixth largest non-Christian denomination in the US, after Judaism (1.4%), Islam (0.6%), Buddhism (0.5%), Hinduism (0.3%) and Unitarian Universalism (0.3%).[35]

The hammer Mjöllnir is one of the primary symbols of Germanic Neopaganism. Pendants of the Mjöllnir are commonly worn amongst Germanic Neopagans.

In Iceland, the members of Ásatrúarfélagið account for 0.4% of the total population,[36] which is just over a thousand people. In Lithuania, many people practice Romuva, a revived version of the pre-Christian religion of that country. Lithuania was among the last areas of Europe to be Christianized.

There are a number of Pagan authors who have examined the relation of the 20th-century movements of polytheistic revival with historical polytheism on one hand and contemporary traditions of indigenous folk religion on the other. Isaac Bonewits introduces a terminology to make this distinction,[37]

- Paleopaganism: A retronym coined to contrast with "Neopaganism", "original polytheistic, nature-centered faiths", such as the pre-Hellenistic Greek and pre-imperial Roman religion, pre-Migration period Germanic paganism as described by Tacitus, or Celtic polytheism as described by Julius Caesar. Among extant "major religions", Bonewits would count as Paleopagan Hinduism as it stood prior to the Islamic invasions of India, Shintoism and Taoism.
- Mesopaganism: A group, which is, or has been, significantly influenced by monotheistic, dualistic, or nontheistic worldviews, but has been able to maintain an independence of religious practices. This group includes aboriginal Americans as well as Australian aborigines, Viking Age Norse paganism and New Age spirituality. Influences include: Freemasonry, Rosicrucianism, Theosophy, Spiritualism, and the many Afro-Diasporic faiths like Haitian Vodou, Santería and Espiritu religion. Isaac Bonewits includes British Traditional Wicca in this subdivision.
- Neopaganism: A movement by modern people to revive nature-worshipping, pre-Christian religions, or other nature-based spiritual paths. This definition may include groups such as Neo-druidism and Wicca at the other.

Prudence Jones and Nigel Pennick in their *A History of Pagan Europe* (1995) classify "pagan religions" as characterized by the following traits:

- polytheism: Pagan religions recognise a plurality of divine beings, which may or may not be considered aspects of an underlying unity (the soft and hard polytheism distinction)
- "nature-based": Pagan religions have a concept of the divinity of Nature, which they view as a manifestation of the divine, not as the "fallen" creation found in Dualistic cosmology.
- "sacred feminine": Pagan religions recognize "the female divine principle", identified as "the Goddess" (as opposed to individual goddesses) besides or in place of the male divine principle as expressed in the Abrahamic God.[38]

### Demographics

Paganism has been previously defined broadly, to encompass many or most of the faith traditions outside the Abrahamic religions.

The term has also been used more narrowly,[39] [40] [41] however, to refer only to religions outside the very large group of so-called Axial Age faiths that encompass both the Abrahamic religions and the chief Indian religions. Under this narrower definition, which differs from that historically used by many[42] [43] (though by no means all[44] [45] ) Christians and other Westerners, contemporary Paganism is a smaller and more marginal numerical phenomenon. According to Encyclopædia Britannica estimates (as of 2005), adherents of Chinese folk religion account for some 6.3% of world population, and adherents of tribal religions ("ethnoreligionists") for another 4.0%. The number of adherents of neopaganism is insignificant in comparison, amounting to 0.02% of world population at the most, or some 0.4% of the "ethnoreligious" population.

## See also

- Animism
- Folk religion
- Ethnic religion
- List of Pagans
- Myth and ritual
- Neopaganism
- Polytheism
- Barbarian
- Kafir

## References

### Notes

[1] http://encarta.msn.com/dictionary_/pagan.html

[2] "CATHOLIC ENCYCLOPEDIA: Paganism" (http://www.newadvent.org/cathen/11388a.htm). 21 November 2009. . Retrieved 17 August 2010. "Paganism, in the broadest sense includes all religions other than the true one revealed by God, and, in a narrower sense, all except Christianity, Judaism, and Mohammedanism. The term is also used as the equivalent of Polytheism."

[3] http://www.religioustolerance.org/paganism.htm – Robinson, B.A (2000). "What do "Paganism" & "Pagan" mean?" at religioustolerance.org

[4] "Pagan", Encyclopædia Britannica 11th Edition, 1911, retrieved 22 May 2007. (http://encyclopedia.jrank.org/ORC_PAI/PAGAN_Lat_paganus_of_or_belongi.html)

[5] Peter Brown, in Glen Warren Bowersock, Peter Robert Lamont Brown, Oleg Grabar, eds., *Late Antiquity: a guide to the postclassical world*, 1999, *s.v.* "Pagan".

[6] "A Basic Introduction to Paganism" (http://www.bbc.co.uk/dna/h2g2/classic/A1032166), BBC, retrieved 19 May 2007.

[7] http://encarta.msn.com/dictionary_/pagan.html Word History

[8] Watts, Alan W. "Nature, Man and Woman", 1991, Vintage Books, p. 25.

[9] http://dictionary.oed.com

[10] OED etymology for paynim: < Anglo-Norman paenisme, painisme, paienime, painnim, peinibvgnb, bjkbyh ikbh where is the pegacorn me, paenime, etc., and Old French paienime, paienisme heathen lands (c1150–74), heathen religion (1160) < post-classical Latin paganismus (see PAGANISM n.), probably influenced by Old French paien (see PAYEN n.).

[11] OED entry for 'paynimry'.

[12] http://www.lexic.us/definition-of/paynimry

[13] http://www.perseus.tufts.edu/cgi-bin/ptext?doc=Perseus%3atext%3a1999%2e04%2e0062&query=id%3dpagus#id,pagus Harry Thurston Peck, *Harper's Dictionary of Classical Antiquity*, 1897; "pagus"

[14] Divers. Quaest. 83. Augustine makes clear that, in his time, *paganus* was the term in Vulgar Latin synonymous to educated *gentilis* "gentile".

[15] "Theodosius I", The Catholic Encyclopedia, 1912 (http://www.newadvent.org/cathen/14577d.htm)

[16] "The City of God", Britannica Ultimate Reference Suite DVD, 2003.

[17] 'Hymn to Proserpine'
[18] Jewish Encyclopedia (http://www.jewishencyclopedia.com/view.jsp?artid=489&letter=C&search=polytheism christianity)
[19] Shirk
[20] Chapman, John (1912). "Tritheists" (http://www.newadvent.org/cathen/15061b.htm), *The Catholic Encyclopedia*, Vol. 15. New York: Robert Appleton Company. Retrieved 17 May 2011.
[21] Christianised calendar
[22] The Pope, The Emperor and the Persian Leader (http://asharqalawsat.com/english/news.asp?section=2&id=6465)
[23] 'Philip Melanchthon 'Apologia Confessionis Augustanae'
[24] Jean Seznec 'The Survival of the Pagan Gods'
[25] c.f. the civil, natural and mythical theologies of Marcus Terentius Varro
[26] Ludwig Feuerbach, *Geschichte der neueren Philosophie* (1833), Introduction, §1 (Paganism, Philosophy, Religion, Christianity) *Das Wesen des Heidentums war die Einheit von Religion und Politik, Geist und Natur, Gott und Mensch. Aber der Mensch im Heidentum war nicht der Mensch schlechtweg, sondern der nationell bestimmte Mensch: der Grieche, der Römer, der Ägyptier, der Jude, folglich auch sein Gott ein nationell bestimmtes, besonderes, dem Wesen oder Gotte anderer Völker entgegengesetztes Wesen — ein Wesen also im Widerspruch mit dem Geiste, welcher das Wesen der Menschheit und als ihr Wesen die allgemeine Einheit aller Völker und Menschen ist. Die Aufhebung dieses Widerspruchs im Heidentum war die heidnische Philosophie; denn sie riß den Menschen heraus aus seiner nationellen Abgeschlossenheit und Selbstgenügsamkeit, erhob ihn über die Borniertheit des Volksdünkels und Volksglaubens, versetzte ihn auf den kosmopolitischen Standpunkt.*
[27] E. Kessler, *Dionysian Monotheism in Nea Paphos, Cyprus* "two monotheistic religions, Dionysian and Christian, existed contemporaneously in Nea Paphos during the 4th century C.E. [...] the particular iconography of Hermes and Dionysos in the panel of the Epiphany of Dionysos [...] represents the culmination of a Pagan iconographic tradition in which an infant divinity is seated on the lap of another divine figure; this Pagan motif was appropriated by early Christian artists and developed into the standardized icon of the Virgin and Child. Thus the mosaic helps to substantiate the existence of Pagan monotheism." Biblical Studies on the Web (http://www.bsw.org/?l=71851&a=Comm06.html)
[28] Pausanias, *Description of Greece* 6. 26. 1 - 2
[29] Athenaeus, *Deipnosophistae* 2. 34a
[30] Wick, Peter (2004). "Jesus gegen Dionysos? Ein Beitrag zur Kontextualisierung des Johannesevangeliums" (http://www.bsw.org/?l=71851&a=Comm06.html). *Biblica* (Rome: Pontifical Biblical Institute) **85** (2): 179–198. . Retrieved 2007-10-10.
[31] Studies in Early Christology (http://books.google.com/books?id=0fLPOx1B-AwC&pg=PA331&lpg=PA331&dq="dionysus+had+been+at+home+in+palestine+for+a+long+time"&source=web&ots=GHsCkhiNP6&sig=qE6Sov5Xi_LB_zpRAQZreSAekTQ), by Martin Hengel, 2005, p.331 (ISBN 0567042804)
[32] Powell, Barry B., *Classical Myth* Second ed. With new translations of ancient texts by Herbert M. Howe. Upper Saddle River, New Jersey: Prentice-Hall, Inc., 1998.
[33] "It would be a great pleasure to make the comparison with what survives to us of ancient Paganism in our old books, in order to have better [grasped] their spirit." Peter N. Miller, *History of Religion Becomes Ethnology: Some Evidence from Peiresc's Africa* Journal of the History of Ideas 67.4 (2006) 675–696. (http://muse.jhu.edu/journals/journal_of_the_history_of_ideas/v067/67.4miller.html)
[34] Jack Zipes, *The Great Fairy Tale Tradition: From Straparola and Basile to the Brothers Grimm*, p 846, ISBN 0-393-97636-X
[35] ARIS 2001 figures.
[36] Statistics Iceland – Statistics >> Population >> Religious organisations (http://www.statice.is/?PageID=1180&src=/temp_en/Dialog/varval.asp?ma=MAN10001&ti=Populations+by+religious+organizations+1990-2008+&path=../Database/mannfjoldi/Trufelog/&lang=1&units=Number)
[37] "Defining Paganism: Paleo-, Meso-, and Neo-" (http://www.neopagan.net/PaganDefs.html) (Version 2.5.1) 1979, 2007 c.e., Isaac Bonewits
[38] Jones, Prudence; Pennick, Nigel (1995). *A History of Pagan Europe*. Page 2. Routledge.
[39] Meanings of the terms Pagan and Paganism (http://www.religioustolerance.org/paganism.htm)
[40] Eisenstadt, S.N., 1983, Transcendental Visions – Other-Worldliness – and Its Transformations: Some More Comments on L. Dumont. *Religion*13:1–17, at p. 3.
[41] Michael York, Paganism as Root-Religion, The Pomegranate, 6:1 (2004), pp. 11–18 (distinguishing the main streams of developed religion as gnostic, dharmic, Abrahamic and pagan).
[42] Catholic Encyclopaedia (1917 edition) on paganism (http://www.newadvent.org/cathen/11388a.htm)
[43] Hindu rites at a famous Catholic shrine shocks many Catholics (http://www.cwnews.com/news/viewstory.cfm?recnum=33161)
[44] David Scott, Christian Responses to Buddhism in Pre-Medieval Times, Numen, Vol. 32, No. 1 (Jul., 1985), pp. 88–100
[45] Audrius Beinorius, Buddhism in the Early European Imagination: A Historical Perspective, ACTA ORIENTALIA VILNENSIA 6:2 (2005), pp. 7–22

### Bibliography

- Robert, P. & Scott, N., (1995) "*A History of Pagan Europe*". New York, Barnes & Noble Books, ISBN 0-7607-1210-7.
- York, Michael *Pagan Theology: Paganism as a World Religion* NYU Press (2003), ISBN 0814797083.


# Pantheism

**Pantheism** is the view that the Universe (Nature) and God (or divinity) are identical.[1] Pantheists thus do not believe in a personal, anthropomorphic or creator god. The word derives from the Greek (*pan*) meaning "all" and the Greek (*theos*) meaning "God". As such, Pantheism denotes the idea that "God" is best seen as a process of relating to the Universe.[2] Although there are divergences within Pantheism, the central ideas found in almost all versions are the Cosmos as an all-encompassing unity and the sacredness of Nature.

In Pantheism, God is identical with the universe, but in Pan*en*theism God lies within and also beyond or outside of the universe.[5]

There is no official universal symbol for all types of Pantheism, but one icon is used by the World Pantheist Movement (WPM): the spiral as seen in the curves of the nautilus shell, or in the spiral arms of a galaxy, showing the link between the cosmic physical and the biological. The spiral represents a variety of things: it means evolution, eternity, spirituality, growth.[3] Sometimes the Nautilus spiral alone is used; it embodies the Fibonacci series and the golden ratio.[4]

## History

The term "Pantheism" is derived from Greek words *pan* (Ancient Greek: πᾶν) meaning "all" and *theos* (*θεός*) meaning God, in the sense of theism. The term *pantheist* — from which the word *Pantheism* was derived — was purportedly first used in English by Irish writer John Toland in his 1705 work "Socinianism Truly Stated, by a pantheist". He clarified the idea in a 1710 letter to Gottfried Leibniz when he referred to "the pantheistic opinion of those who believe in no other eternal being but the universe".[6] However, many earlier writers, schools of philosophy, and religious movements expressed pantheistic ideas.

They include some of the Presocratics, such as Heraclitus and Anaximander. The Stoics were Pantheists, beginning with Zeno of Citium and culminating in the emperor-philosopher Marcus Aurelius. During the pre-Christian Roman Empire, Stoicism was one of the three dominant schools of philosophy, along with Epicureanism and Neoplatonism. The early Taoism of Lao Zi and Zhuangzi is also sometimes considered pantheistic.[6]

In the West, Pantheism went into retreat during the Christian years between the 4th and 15th centuries, when it was regarded as heresy. The first open revival was by Giordano Bruno (burned at the stake in 1600). Baruch Spinoza's *Ethics*, finished in 1675, was the major source from which Pantheism spread (though Spinoza himself did not use the word, and there is some controversy over whether he may more accurately be termed a panentheist.[7] John Toland was influenced by both Spinoza and Bruno. In 1720 he wrote the *Pantheisticon: or The Form of Celebrating the Socratic-Society* in Latin.[8]

In 1785 a major controversy known in German as the Pantheismus-Streit (Pantheism controversy) between critic Friedrich Jacobi and defender Moses Mendelssohn helped to spread awareness of Pantheism to many German thinkers in the late 18th and in the 19th century.[9]

For a time during the 19th century Pantheism was the religious viewpoint of many leading writers and philosophers, attracting figures such as William Wordsworth and Samuel Coleridge in Britain; Johann Gottlieb Fichte, Friedrich Wilhelm Joseph Schelling and Georg Wilhelm Friedrich Hegel in Germany; Walt Whitman, Ralph Waldo Emerson and Henry David Thoreau in the USA. Seen as a growing threat by the Vatican, it came under attack in the Syllabus of Errors of Pius IX.[10]

However, in the 20th century Pantheism was sidelined by political ideologies such as Communism and Fascism, by the traumatic upheavals of two world wars, and later by relativistic philosophies such as Existentialism and Post-Modernism. It persisted in eminent pantheists such as the novelist D. H. Lawrence, scientist Albert Einstein, poet Robinson Jeffers, architect Frank Lloyd Wright and historian Arnold Toynbee.[6]

## Recent developments

In the late 20th century, Pantheism began to see a resurgence.[6] Pantheism chimed with the growing ecological awareness in society and the media. It was described as "Hollywood's religion of choice for a generation now",[11] and often declared to be the underlying "theology" of Paganism.[12] 1975 saw the foundation of the Universal Pantheist Society, which remains extremely small. The creation of the naturalistic World Pantheist Movement in 1999, with its multiple mailing lists and social networks, led to much wider visibility.

Richard Dawkins' *The God Delusion* gave Naturalistic Pantheism increased credibility among atheists by describing it as "sexed-up atheism." [13] The Vatican gave Pantheism further prominence in a Papal encyclical of 2009[14] and a New Year's Day statement on January 1, 2010,[15] which criticized Pantheism for denying the superiority of humans over nature and "seeing the source of man's salvation in nature".[14]

In 2008, Albert Einstein's 1954 German letter in which he dismissed belief in God was auctioned off for more than $330,000 US. Einstein wrote, "the word 'God' is for me nothing more than the expression and product of human weaknesses, the Bible a collection of honourable but still primitive legends which are nevertheless pretty childish."[16] "I do not believe in a personal God and I have never denied this but have expressed it clearly," he wrote in another letter in 1954. Einstein relates his belief to Pantheism: "If something is in me which can be called religious then it is the unbounded admiration for the structure of the world so far as our science can reveal it."[17]

## Varieties

All varieties of Pantheism involve reverence for the Universe/Cosmos as a totality, and all stress some kind of unity. All have a strong emphasis on the natural world as a focus of reverence and of ethics. There are three major categories of Pantheism, which differ as to whether they regard reality as made up of only one type of substance, or two, and what that type of substance is.[6]

### Monist physicalist pantheism

Monist physicalist Pantheism or Naturalistic Pantheism holds that there is only one type of substance, and that substance is physical, i.e. able at its most basic level to be described by physics, though more complex phenomena such as life, consciousness and societies can appear through emergence. Historically this version was held by Stoics such as Zeno of Citium or Marcus Aurelius, and in modern times by John Toland, Ernst Haeckel, D.H. Lawrence and Paul Harrison. This version is represented today by the World Pantheist Movement. In this version, the term *god* — if used at all — is basically a synonym for Nature or Universe, seen from the point of view of reverence.

### Monist idealist pantheism

Monist idealist Pantheism or Monistic Idealism holds that there is only one type of substance, and that substance is mental or spiritual. Some versions hold that the ultimate reality consists of a single cosmic consciousness. This version is common in Hindu philosophies and Consciousness-Only schools of Buddhism, as well as in some New Age writers such as Deepak Chopra. This is distinguished from pandeism in that pandeism asserts that the whole of reality was at some time sentient.

### Dualist pantheism

Dualist Pantheism holds that there are two major types of substance, physical and mental/spiritual, which interact or are unified in some way. Dualistic pantheism is very diverse, and may include beliefs in reincarnation, cosmic consciousness, and paranormal connections across Nature. Criticisms of this interpretation are generally related to the Mind-body problem.[18]

## Issues

### Use of religious vocabulary

A significant debate within the pantheistic community is about the use of the word "God." Pantheists do not believe in a God in the common and traditional sense of a personal creator being. Some modern Pantheists avoid using God-words altogether, since they regard them as misleading. Others feel that the word God is essential to express the strength of their feelings towards Nature and the Universe.

Some critics have argued that pantheism is little more than a redefinition of the word "God" to mean "Nature," "Universe", or "reality".

When pantheism is considered in relation to theism, there is a denial of theistic claims. For example, theism is the belief in a "personal" God that transcends (is separate from) the world. Pantheists deny the existence of a personal God. Some deny the existence of a Being that has intentional states and associated capacities such as the ability to make decisions. Some Pantheists recognize the intelligence involved in the rational and systematic functioning of Nature and the Universe, which is not necessarily intentional but is "minded" in a sense. Pandeism differs from Pantheism in that pandeism leaves that possibility of a sentient deity before the creation of the universe.

There are disagreements as to whether Pantheism is atheistic or not. Atheists argue the non-theistic god of pantheism is not a god (according to the traditional definition),[19] while others suggest a deity is not necessarily transcendent.[20]

## Similar concepts in other religious traditions

### Taoism

Taoism is pantheistic at least in the writings of its leading thinkers Lao Tzu and Zhuangzi, although it later developed into a folk religion with many deities.
The Tao te Ching by Lao Tzu rarely speaks of a personal or creator God. Its central focus, the Tao or Way, is conceived of as a mysterious and numinous unity, infinite and eternal, underlying all things and sustaining them. The Tao is always spoken of with profound religious reverence and respect, similar to the way that Pantheism discusses the "divinity" of the Universe.[6] The ideal of Taoism was to live in harmony with the Tao and to cultivate a simple and frugal life, avoiding unnecessary action: "Being one with nature, he [the sage] is in accord with the Tao."[21]

Zhuangzi emphasized the pantheistic content of Taoism even more clearly. "Heaven and I were created together, and all things and I are one." When Tung Kuo Tzu asked Zhuangzi where the Tao was, he replied that it was in the ant, the grass, the clay tile, even in excrement: "There is nowhere where it is not... There is not a single thing without Tao."[22]

### Hinduism

It is generally asserted that Hindu religious texts are the oldest known literature that contains Pantheistic ideas.[23] In Hindu theology, Brahman is the unchanging, infinite, immanent, and transcendent reality which is the Divine Ground of all things in this Universe, and is also the sum total of all that ever is, was, or ever shall be. This idea of pantheism is traceable from some of the more ancient Vedas and Upanishads to vishishtadvaita philosophy. All Mahāvākyas (Great Sayings) of the Upanishads, in one way or another, seem to indicate the unity of the world with the Brahmam. It further says "This whole universe is Brahman, from Brahman to a clod of earth." Pantheism is a key component of Advaita philosophy. Other subdivisions of Vedanta do not strictly hold this tenet.

### Other religions

There are many elements of pantheism in some forms of Buddhism, Sufism, Sikhism, Neopaganism, and Theosophy as well as in several tendencies in the major theistic religions. See also the Neopagan section of Gaia and the Church of All Worlds.

Many Unitarian Universalists consider themselves pantheists. The Islamic religious tradition, in particular Sufism and Alevism has a strong belief in the unitary nature of the universe and the concept that everything in it is an aspect of God itself, although this perspective leans closer to panentheism and may also be termed Theopanism. Many traditional and folk religions including African traditional religions and Native American religions can be seen as pantheistic, or a mixture of pantheism and other doctrines such as polytheism and animism.

## Distinction from related concepts

Some other theological models have attempted to incorporate the perceived benefits of pantheism with the perceived benefits of classical monotheism.

The term panentheism (from Greek πᾶν (pân) "all"; ἐν (en) "in"; and θεός (theós) "God"; "all-in-God") was formally coined in Germany in the 19th Century to express a philosophical synthesis between traditional theism and pantheism, that God is substantially omnipresent in the physical universe but also in a sense exists "apart from" or "beyond" the universe as its Creator and Sustainer.[24] Thus panentheism is not compatible with pantheism, in which God and the universe are synonymous—with no part of God considered as being distinct from the universe.[25] [26]

For the same reasons, pandeism is *not* a form of pantheism. Though pandeism is characterized as a combination of reconcilable elements of pantheism and deism,[27] [28] it is simply a form of deism which uses some pantheistic terminology while still including a Creator-deity which is at some point distinct from the universe.

## See also

- List of Pantheists
- Monism
- Naturalistic pantheism
- Nature worship (disambiguation)
- Universal Pantheist Society
- World Pantheist Movement

## References


[1] *The New Oxford Dictionary Of English.* Oxford: Clarendon Press. 1998. pp. 1341. ISBN 0-19-861263-X.

[2] Owen, H. P. *Concepts of Deity.* London: Macmillan, 1971.

[3] pantheism.net (http://www.pantheism.net).

[4] pantheism.net (http://www.pantheism.net/pan/Pan9.pdf).

[5] John Culp (May 19, 2009). "Panentheism" (http://plato.stanford.edu/archives/sum2009/entries/panentheism). *The Stanford Encyclopedia of Philosophy.* Stanford, CA page=: Stanford Uni versity. . "Peacocke identifies his understanding of God's relation to the world as panentheism because of its rejection of dualism and external interactions by God in favor of God always working from inside the universe. At the same time, God transcends the universe because God is infinitely more than the universe".

[6] Paul Harrison, *Elements of Pantheism,* 1999.

[7] Genevieve Lloyd, Routledge Philosophy GuideBook to Spinoza and The Ethics (Routledge Philosophy Guidebooks), Routledge; 1 edition (October 2, 1996), ISBN 978-0-415-10782-2, Page: 40

[8] Toland: The father of modern pantheism (pantheism.net) (http://www.pantheism.net/paul/toland.htm).

[9] Friedrich Heinrich Jacobi, in Stanford Encyclopedia of Philosophy (plato.stanford.edu) (http://plato.stanford.edu/entries/friedrich-jacobi/ ).

[10] Syllabus or Errors 1.1 (papalencyclicals.net) (http://www.papalencyclicals.net/Pius09/p9syll.htm).

[11] Heaven and Nature, Ross Douthat, New York Times, December 20, 2009.

[12] Margot Adler, Drawing Down the Moon, Beacon Press, 1986.

[13] The God Delusion. Boston: Houghton Mifflin. 2006.

[14] Caritas In Veritate, July 7, 2009.

[15] Message Of His Holiness Pope Benedict XVI For The Celebration Of The World Day Of Peace.

[16] "Einstein letter dismissing God sells for $330,000 US" (http://www.cbc.ca/news/world/story/2008/05/15/einstein-letter.html). CBC Canada. 2008-05-15. . Retrieved 2011-08-31.

[17] "Belief in God a 'product of human weaknesses': Einstein letter" (http://www.cbc.ca/news/world/story/2008/05/13/einstein-religion.html). CBC Canada. 2008-05-13. . Retrieved 2011-08-31.

[18] *Mind & Society,* Volume 1, Number 2, 73-85, DOI: 10.1007/BF02512314.

[19] Dawkins, R (2006), *The God Delusion,* Transworld, a Black Swan Book, ISBN 978-0-552-77331-7 "Pantheism is sexed-up atheism".

[20] "With some exceptions, pantheism is non-theistic, but it is not atheistic." Stanford Encyclopedia of Philosophy, Pantheism (plato.stanford.edu) (http://plato.stanford.edu/entries/pantheism/#Ath).

[21] Tao te Ching, 16.

[22] Chuang Tzu - The butterfly philosopher (pantheism.net) (http://www.pantheism.net/paul/chuang.htm).

[23] *General Sketch of the History of Pantheism,* p. 29.

[24] John W. Cooper, *The Other God of the Philosophers,* Baker Academic, 2006, p. 27.

[25] What is Panentheism? (http://atheism.about.com/library/FAQs/religion/blrel_theism_panen.htm), atheism.about.com. About Agnosticism/Atheism. Retrieved 2 October 2009.

[26] Erwin Fahlbusch, Geoffrey William Bromiley, David B. Barrett (1999). *The Encyclopedia of Christianity* (http://books.google.com/books?id=sCY4sAjTGIYC&pg=PA21). Wm. B. Eerdmans Publishing. pp. 21. ISBN 0802824161. .

[27] Sean F. Johnston (2009). *The History of Science: A Beginner's Guide.* pp. 90. ISBN 1851686819.

[28] Alex Ashman, *BBC News,* " Metaphysical Isms (http://www.bbc.co.uk/dna/h2g2/A25172309)".


## External links

- Beliefs, History and Practice of Pantheism (pantheism.net) (http://www.pantheism.net/paul)
- Stanford Encyclopedia of Philosophy entry (plato.stanford.edu) (http://plato.stanford.edu/entries/pantheism/)
- Pantheism (http://www.dmoz.org/Society/Religion_and_Spirituality/Pantheism//) at the Open Directory Project
- The Pantheist Index (http://www.pantheist-index.net/), pantheist-index.net
- An Introduction to Pantheism (wku.edu) (http://www.wku.edu/~jan.garrett/panthesm.htm)
- "Pantheism". *Catholic Encyclopedia.* New York: Robert Appleton Company. 1913.
- World Pantheist Movement (pantheism.net) (http://www.pantheism.net)

# Animism

**Animism** (from Latin *anima* "soul, life")[1] [2] refers to the belief that non-human entities are spiritual beings, or at least embody some kind of life-principle.[3]

Animism encompasses the beliefs that there is no separation between the spiritual and physical (or material) world, and souls or spirits exist, not only in humans, but also in all other animals, plants, rocks, natural phenomena such as thunder, geographic features such as mountains or rivers, or other entities of the natural environment.[4] Animism may further attribute souls to abstract concepts such as words, true names, or metaphors in mythology. Animism is particularly widely found in the religions of indigenous peoples,[5] including Shinto, and some forms of Hinduism, Buddhism, Pantheism, Paganism, and Neopaganism.

Throughout European history, philosophers such as Aristotle and Thomas Aquinas, among others, contemplated the possibility that souls exist in animals, plants, and people; however, the currently accepted definition of animism was only developed in the 19th century by Sir Edward Tylor, who created it as "one of anthropology's earliest concepts, if not the first".[5]

According to the anthropologist Tim Ingold, animism shares similarities to totemism but differs in its focus on individual spirit beings which help to perpetuate life, whereas totemism more typically holds that there is a primary source, such as the land itself or the ancestors, who provide the basis to life. Certain indigenous religious groups such as the Australian Aborigines are more typically totemic, whereas others like the Inuit are more typically animistic in their worldview.[6]

## Etymology

The term *animism* appears to have been first developed as *animismus* by German scientist Georg Ernst Stahl, circa 1720, to refer to the "doctrine that animal life is produced by an immaterial soul." The actual English language form of *animism*, however, can only be attested to 1819.[7] The term was taken and redefined by the anthropologist Sir Edward Tylor in his 1871 book *Primitive Culture*, in which he defined it as "the general doctrine of souls and other spiritual beings in general."[8] According to religious scholar Robert Segal, Tylor saw all religions, "modern and primitive alike", as forms of animism.[1]

Sir Edward Tylor was responsible for forming the definition of animism currently accepted in anthropology.

According to Tylor, animism often includes "an idea of pervading life and will in nature";[9] i.e., a belief that natural objects other than humans have souls. As a self-described "confirmed scientific rationalist", Tylor believed that this view was "childish" and typical of "cognitive underdevelopment",[10] and that it was therefore common in "primitive" peoples such as those living in hunter gatherer societies.

Tylor's definition of animism has since largely been followed by anthropologists, such as Émile Durkheim, Claude Lévi-Strauss and Tim Ingold. However, some anthropologists, such as Nurit Bird-David, have criticised the Tylorian concept of animism, believing it to be outdated.[11]

## Motivation

Further information: anthropology of religion and psychology of religion

Animism in the widest sense, i.e., thinking of objects as animate, and treating them as if they were animate, is near-universal. Jean Piaget applied the term in child psychology in reference to an implicit understanding of the world in a child's mind which assumes that all events are the product of intention or consciousness. Piaget explains this with a cognitive inability to distinguish the external world from one's own psyche. Developmental psychology has since established that the distinction of animate vs. inanimate things is an abstraction acquired by learning.

The justification for attributing life to objects was stated by David Hume in his *Natural History of Religion* (Section III): "There is a universal tendency among mankind to conceive all beings like themselves, and to transfer to every object those qualities with which they are familiarly acquainted, and of which they are intimately conscious."[12]

Lists of phenomena, from the contemplation of which "the savage" was led to believe in animism, have been given by Sir E. B. Tylor, Herbert Spencer, Andrew Lang, and others; a controversy arose between the former as to the priority of their respective lists. Among these phenomena are trance states, dreams, and hallucinations.

## Animism and religion

Animism is a belief held in many religions around the world, and is not, as some have purported, a type of religion in itself. It is a belief, such as shamanism, polytheism or monotheism, that is found in several religions. In modern usage, the term is sometimes used improperly as a catch-all classification of "other world religions" alongside major organized religions.

### Origin of religion

Some theories have been put forward that the anthropocentric belief in animism among early humans was the basis for the later evolution of religions. In one such theory, put forward by Sir E. B. Tylor, early humans initially, through mere observation, recognized what might be called a soul, life-force, spirit, breath or animus within themselves; that which was present in the body in life and absent in death. These early humans equated this soul with figures which would appear in dreams and visions. These early human cultures later interpreted these spirits to be present in animals, the living plant world, and even in natural objects in a form of animism. Eventually, these early humans grew to believe that the spirits were invested and interested in human life, and performed rituals to propitiate them. These rituals and beliefs eventually evolved over time into the vast array of "developed" religions. According to Tylor, the more scientifically advanced the society, the less that society believed in Animism; however, any remnant ideologies of souls or spirits, to Tylor, represented "survivals" of the original animism of early humanity.{{link [13]}}

### World view

In many animistic world views found in hunter-gatherer cultures, the human being is often regarded as on a roughly equal footing with other animals, plants, and natural forces.[14] Therefore, it is morally imperative to treat these agents with respect. In this world view, humans are considered a part of nature, rather than superior to, or separate from it. In such societies, ritual is considered essential for survival, as it wins the favor of the spirits of one's source of food, shelter, and fertility and wards off malevolent spirits. In more elaborate animistic religions, such as Shinto, there is a greater sense of a special character to humans that sets them apart from the general run of animals and objects, while retaining the necessity of ritual to ensure good luck, favorable harvests, and so on.

## Death

Most animistic belief systems hold that the spirit survives physical death. In some systems, the spirit is believed to pass to an easier world of abundant game or ever-ripe crops, while in other systems, the spirit remains on earth as a ghost, often malignant. Still other systems combine these two beliefs, holding that the soul must journey to the spirit world without becoming lost and thus wandering as a ghost (*e.g.*, the Navajo religion). Funeral, mourning rituals, and ancestor worship performed by those surviving the deceased are often considered necessary for the successful completion of this journey.

From the belief in the survival of the dead arose the practice of offering food, lighting fires, etc., at the grave, at first, maybe, as an act of friendship or filial piety, later as an act of ancestor worship. The simple offering of food or shedding of blood at the grave develops into an elaborate system of sacrifice. Even where ancestor worship is not found, the desire to provide the dead with comforts in the future life may lead to the sacrifice of wives, slaves, animals, and so on, to the breaking or burning of objects at the grave or to the provision of the ferryman's toll: a coin put in the mouth of the corpse to pay the traveling expenses of the soul.

But all is not finished with the passage of the soul to the land of the dead. The soul may return to avenge its death by helping to discover the murderer, or to wreak vengeance for itself. There is a widespread belief that those who die a violent death become malignant spirits and endanger the lives of those who come near the haunted spot. In Malay folklore, the woman who dies in childbirth becomes a pontianak, a vampire-like spirit who threatens the life of human beings. People resort to magical or religious means of repelling spiritual dangers from such malignant spirits.

It is not surprising to find that many peoples respect and even worship animals (see *totem* or *animal worship*), often regarding them as relatives. It is clear that widespread respect was paid to animals as the abode of dead ancestors, and much of the cults to dangerous animals is traceable to this principle; though there is no need to attribute an animistic origin to it.[15]

The practice of head shrinking among Jivaroan and Urarina peoples derives from an animistic belief that if the spirits of one's mortal enemies (i.e., the nemesis of one's being) are not trapped within the head, they can escape slain bodies. After the spirit transmigrates to another body, they can take the form of a predatory animal and even exact revenge.

## Mythology

A large part of mythology is based upon a belief in souls and spirits — that is, upon animism in its more general sense. Urarina myths that portray plants, inanimate objects, and non-human animals as personal beings are examples of animism in its more restrictive sense.[16]

However, many mythologies focus largely on corporeal beings rather than "spiritual" ones; the latter may even be entirely absent. Stories of transformation, deluge and doom myths, and myths of the origin of death do not necessarily have any animistic basis.

As mythology began to include more numerous and complex ideas about a future life and purely spiritual beings, the overlap between mythology and animism widened. However, a rich mythology does not necessarily depend on a belief in many spiritual beings.

## Philosophy

The term "animism" has been applied to many different philosophical systems. It is used to describe Aristotle's view of the relation of soul and body held also by the Stoics and Scholastics. On the other hand monadology (Leibniz) has also been termed animistic. The name is most commonly applied to vitalism, a view mainly associated with Georg Ernst Stahl and revived by Francisque Bouillier (1813–1899), which makes life, or life and mind, the directive principle in evolution and growth, holding that all cannot be traced back to chemical and mechanical processes, but that there is a directive force which guides energy without altering its amount. An entirely different class of ideas, also termed animistic, is the belief in the world soul (anima mundi), held by Plato, Schelling and others.

### Distinction from Pantheism

Animism is not the same as Pantheism, although the two are sometimes confused. Some faiths and religions are even both pantheistic and animistic. One of the main differences is that while animists believe everything to be spiritual in nature, they do not necessarily see the spiritual nature of everything in existence as being united (monism), the way pantheists do. As a result, animism puts more emphasis on the uniqueness of each individual soul. In Pantheism, everything shares the same spiritual essence, rather than having distinct spirits and/or souls.

## Contemporary animist traditions

Further information: Folk religion and Shamanism

- African traditional religions, a group of beliefs in various spirits of nature, are not commonly described as animistic, yet this fact has for many years been disputed by leading cultural anthropologists.
- In the Canary Islands (Spain), aboriginal Guanches professed an animistic religion.
- Shinto, the traditional religion of Japan, is highly animistic. In Shinto, spirits of nature, or kami, are believed to exist everywhere, from the major (such as the goddess of the sun), which can be considered polytheistic, to the minor, which are more likely to be seen as a form of animism.
- Many traditional beliefs in the Philippines centuries ago, and perhaps still practiced by some to an extent or in seclusion today, are beliefs strongly associated with animism and spiritual beliefs, which have featured rituals aimed at pacifying malevolent spirits.
- There are some Hindu groups which may be considered animist. The coastal Karnataka has a different tradition of praying to spirits (see also Folk Hinduism).
- Many traditional Native American religions are fundamentally animistic. See, for example, the Lakota Sioux prayer Mitakuye Oyasin. The Haudenausaunee Thanksgiving Address, which can take an hour to recite, directs thanks towards every being - plant, animal and other.
- The New Age movement commonly purports animism in the form of the existence of nature spirits.
- Modern Neopagans, especially Eco-Pagans, sometimes like to describe themselves as animists, meaning that they respect the diverse community of living beings and spirits with whom humans share the world/cosmos.[17]
- New tribalist American author Daniel Quinn identifies himself as an animist and defines animism not as a religious belief but a religion itself,[18] though with no holy scripture, organized institutions, or established dogma.[19] He considers animism the first worldwide religion, common among all tribal societies before the proliferation of "Taker" culture along with this culture's organized, "salvationist" religions.[20] His first discussions of animism appear in his two 1994 books: his novel, *The Story of B*, and autobiography, *Providence: The Story of a Fifty-Year Vision Quest.*

## See also

General

- Anecdotal cognitivism
- Animatism
- Celtic polytheism
- Ethereal being
- Folk religion
- Hyang (Indonesian mythology)
- Hylozoism
- Mana
- Nature worship

- Panpsychism
- Ryukyuan Shinto
- Shamanism in Siberia
- Shamanism
- Shinto
- Tree worship
- Wildlife totemization
- Witchcraft

Related

- Monotheism
- Pantheism
- Polytheism
- Tengriism
- Theism

# References

## Notes

[1] Segal, p. 14.
[2] "Animism", *The American Heritage Dictionary of the English Language*, p. 72.
[3] Iannone, A. Pablo (2001). "animism" (http://books.google.com/books?id=7wBmBO3vpE4C&pg=PA54). *Dictionary of world philosophy*. Taylor & Francis. p. 54. ISBN 9780415179959. .
[4] "The concept that humans possess souls and that souls have life apart from human bodies before and after death are central to animism, along with the ideas that animals, plants, and celestial bodies have spirits" (Wenner).
[5] Bird-David, 1999: p. 67.
[6] Ingold, Tim. (2000). "Totemism, Animism and the Depiction of Animals" in *The Perception of the Environment: Essays on Livelihood, Dwelling and Skill*. London: Routledge, pp. 112-113.
[7] Harper, Douglas. (2001). http://www.etymonline.com/index.php?term=animism
[8] Tylor, E.B. Primitive Culture (http://books.google.com/books?ei=ZHSWTJCnEcL38AbutNWMDA&ct=result&output=text&id=AucLAAAAIAAJ&dq=primitive+culture&q=doctine+of+souls#v=snippet&q="the doctrine of souls and other spiritual beings in general"&f=false). London: John Murray, 1871. p. 21.
[9] Tylor, E.B. Primitive Culture (http://books.google.com/books?ei=ZHSWTJCnEcL38AbutNWMDA&ct=result&output=text&id=AucLAAAAIAAJ&dq=primitive+culture&q=doctine+of+souls#v=snippet&q="the doctrine of souls and other spiritual beings in general"&f=false). London: John Murray, 1871. p. 260.
[10] Bird-David, 1999: pp. 67-68
[11] Bird-David, 1999:
[12] The Natural History of Religion. D. Hume. p. xix (http://books.google.ca/books?id=zp97KA1prvUC&pg=PR19&lpg=PR19&dq="There+is+an+universal+tendency+among+mankind+to+conceive+all+beings+like+themselves,+and+to+transfer+to+every+object+those+qualities+with+which+they+are+familiarly+acquainted,+and+of+which+they+are+intimately+conscious"&source=web&ots=HOqvUgffdx&sig=uFjv8xMCN1l3BpJCsDFD5Bzylcg&hl=en&sa=X&oi=book_result&resnum=1&ct=result#PPR20,M1).
[13] http://www.britannica.com/EBchecked/topic/611503/Sir-Edward-Burnett-Tylor
[14] Fernandez-Armesto, p. 138.
[15] "Animism", *Encyclopædia Britannica*.
[16] Dean, Bartholomew 2009 *Urarina Society, Cosmology, and History in Peruvian Amazonia*, Gainesville: University Press of Florida ISBN 978-081303378 (http://www.upf.com/book.asp?id=DEANXS07).
[17] Higginbotham, p. ??
[18] Ishmael.org: Q and A #75. (1997). http://www.ishmael.org/Interaction/QandA/Detail.CFM?Record=75l
[19] Ishmael.org: Q and A #400. http://www.ishmael.org/Interaction/QandA/Detail.CFM?Record=400
[20] Quinn, Daniel. *The Story of B*. New York: Bantam Books. 1997

## Bibliography


- Adler, Margot. *Drawing Down the Moon: Witches, Druids, Goddess-Worshippers, and Other Pagans in America.* Penguin, 2006.
- "Animism". *The Columbia Encyclopedia.* 6th ed. 2001-07. *Bartleby.com.* Bartleby.com Inc. (10 July 2008) (http://www.bartleby.com/65/an/animism.html).
- Armstrong, Karen. *A History of God: The 4,000-Year Quest of Judaism, Christianity and Islam.* Ballantine Books, 1994.
- Bird-David, Nurit. 1991. "Animism Revisited: Personhood, environment, and relational epistemology", *Current Anthropology* 40, pp. 67–91. Reprinted in Graham Harvey (ed.) 2002. *Readings in Indigenous Religions* (http://books.google.com/books?id=PaLyYm7kFsAC&pg=PR1&dq=Readings+in+Indigenous+Religions&sig=JbLAs8L73O17IWhb22kasr5bCek#PPR1,M1) (London and New York: Continuum) pp. 72–105.
- Cunningham, Scott. *Living Wicca: A Further Guide for the Solitary Practitioner.* Llewellyn, 2002. --
- Dean, Bartholomew 2009 *Urarina Society, Cosmology, and History in Peruvian Amazonia*, Gainesville: University Press of Florida ISBN 978-0813033785, (http://www.upf.com/book.asp?id=DEANXS07).
- Fernandez-Armesto, Felipe. *Ideas that Changed the World.* Dorling Kindersley, 2003.
- Higginbotham, Joyce (2002). *Paganism: An Introduction to Earth- Centered Religions'.* Llewellyn.
- Segal, Robert (2004). *Myth: A Very Short Introduction.* Oxford University Press.
- Wenner, Sara. "Basic Beliefs of Animism". *Emuseum.* 2001. Minnesota State University, (10 July 2008) (http://www.mnsu.edu/emuseum/cultural/religion/animism/beliefs.html).


# Further reading


- Hallowell, A. Irving. "Ojibwa ontology, behavior, and world view" in Stanley Diamond (ed.) 1960. *Culture in History* (New York: Columbia University Press). Reprinted in Graham Harvey (ed.) 2002. *Readings in Indigenous Religions* (London and New York: Continuum) pp. 17–49.
- Harvey, Graham. 2005. *Animism: Respecting the Living World* (London: Hurst and co.; New York: Columbia University Press; Adelaide: Wakefield Press).
- Ingold, Tim: 'Rethinking the animate, re-animating thought'. Ethnos, 71(1) / 2006: pp. 9–20.
- Wundt, W. (1906). *Mythus und Religion*, Teil II. Leipzig 1906 (*Völkerpsychologie*, volume II).
- Quinn, Daniel. *The Story of B*
- Käser, Lothar: *Animismus. Eine Einführung in die begrifflichen Grundlagen des Welt- und Menschenbildes traditionaler (ethnischer) Gesellschaften für Entwicklungshelfer und kirchliche Mitarbeiter in Übersee.* Liebenzeller Mission, Bad Liebenzell 2004, ISBN 3-921113-61-X.
  - mit dem verkürzten Untertitel *Einführung in seine begrifflichen Grundlagen* auch bei: Erlanger Verlag für Mission und Okumene, Neuendettelsau 2004, ISBN 3-87214-609-2.
- Badenberg, Robert: "How about 'Animism'? An Inquiry beyond Label and Legacy". In: *Mission als Kommunikation. Festschrift für Ursula Wiesemann zu ihrem 75.Geburtstag*, edited by Klaus W. Müller. VTR, Nürnberg 2007; ISBN 978-3-937965-75-8 and VKW, Bonn 2007; ISBN 978-3-938116-33-3.

## External links

- Animism, Rinri, Modernization; the Base of Japanese Robotics (http://www.roboethics.org/icra07/contributions/KITANO Animism Rinri Modernization the Base of Japanese Robo.pdf)
- Urban Legends Reference Pages: Weight of the Soul (http://www.snopes.com/religion/soulweight.asp) (http://www.ishmael.org/Interaction/QandA/list.cfm)
- Return to the Land of Souls. A documentary about animism in Ivory Coast by Jordi Esteva (http://www.returntothelandofsouls.com/)
- Animist Network (http://www.animist-network.de/index_e.html)

# Natural World (TV series)

| *Natural World* | |
|---|---|
|   2010-2011 series title card | |
| **Also known as** | *The Natural World*<br>*The World About Us* |
| **Genre** | Nature documentary |
| **Narrated by** | Various |
| **Country of origin** | United Kingdom |
| **Language(s)** | English |
| **No. of episodes** | >400 |
| **Production** | |
| **Producer(s)** | Various |
| **Editor(s)** | Tim Martin (series editor) |
| **Running time** | 50/60 minutes |
| **Production company(s)** | BBC Natural History Unit |
| **Broadcast** | |
| **Original channel** | BBC Two |
| **Picture format** | Film<br>SD: 576i<br>HD: 1080i |
| **Audio format** | Monaural, Stereo |
| **Original run** | 10 December 1967 – present |
| **External links** | |
| Website [1] | |

***Natural World*** is the longest-running nature documentary series on British television. 2008 marked the series' 25th anniversary under its present title, though its origins can be traced back to its predecessor ***The World About Us*** which began over 40 years ago. *Natural World* is broadcast on BBC Two in the autumn and spring seasons; currently, 14 episodes are commissioned by the BBC each year. Production duties are shared between its in-house Natural History Unit, independent UK production companies and leading wildlife filmmakers from around the world. All episodes are now shot in high definition, and more than 400 have been screened in its first 25 years[2] . The current series editor is Tim Martin.

The latest series of *Natural World* began transmission in July 2010 with a study of harpy eagles in the Orinoco rainforest. Other episodes in the series feature British butterflies, a family of sea otters in Monterey harbour and a tribute to elephant matriarch Echo.

## History

In 1967, colour television was gradually being introduced to British audiences on the BBC's recently launched second channel. David Attenborough, at that time the Controller of BBC Two, was seeking to expand the range of colour programmes on the fledgling channel. Formerly head of the BBC's Travel and Exploration Unit in London, he realised that many of its telecine films had been shot in colour. Attenborough commissioned *The World About Us*, a documentary series with a broad remit of geography, anthropology and natural history subjects. It was described by Barry Paine, a producer and narrator of many of the early programmes, as "a series designed to sell colour television sets"[3] . It was no accident that early episodes featured some of the most vibrant colours in the natural world. The first programme was a film by the French vulcanologist Haroun Tazieff, called simply "Volcano". It was broadcast on Sunday 3 December 1967, which also marked the first full evening of colour television in Britain. Another early episode featured the brilliantly-coloured scarlet ibis.

Due to the difficulty in sourcing colour films, production duties were initially shared between the Travel and Exploration Unit in London and the Natural History Unit in Bristol. Over time, the London contribution dried up and the focus became exclusively on natural history.[4] To reflect this, the series title was altered to *The Natural World* in 1983 and then shortened to its current form in 2003. The first episode under the new title was "Save the Panda", broadcast on 30 October 1983. In September 2008 the BBC announced that the series had been re-commissioned for a further three years.[5]

> I have no doubt that Natural World is not only the doyen and founding member of the 50-minute natural history genre but is still the one with the best and most distinguished record.
>
> — David Attenborough

## Format and content

In commissioning *The World About Us*, Attenborough also created the first documentary series whose programmes ran to 50 minutes rather than the half-hour films typical of the time. This allowed wildlife filmmakers to present more detailed studies of their subjects.[6] The format has remained largely unchanged to the present day, though some recent episodes have been extended to 60 minutes.

Although the series is characterised by large budgets and high production values, *Natural World* programmes rarely fall into the category of blue-chip wildlife films (as defined by Derek Bouse[7] ). The commissioning editors draw on a diverse range of films with sometimes idiosyncratic presentation styles or subject matters. Films with conservation messages, human-animal interaction and poetic or spiritual musings on the natural world feature alongside more traditional profiles of individual species and wildlife-rich locations. Some of the more unusual subjects have included plankton, wasps and cephalopods.

## Notable episodes

David Attenborough has maintained a close association with the series over its long history, narrating or presenting around 50 programmes. He narrated the *Echo of the Elephants* films, which followed scientist Cynthia Moss's and cameraman Martyn Colbeck's lengthy study of an African elephant herd in Kenya's Amboseli National Park. In 1996, "Attenborough in Paradise" saw him fulfil a lifelong ambition to observe and film the courtship displays of New Guinea's birds of paradise; he returned to the subject for the 2010 programme "Birds of Paradise".

During the 1980s and 1990s, the *Natural World* strand included occasional mini-series on particular themes. Notable examples include the three-part series *The Flight of the Condor* (1982), *Kingdom of the Ice Bear* (1985) (both released on VHS), and *Wild Indonesia* (1999).

The most requests ever for an episode to be repeated were for *My Halcyon River* by Philippa Forrester and Charlie Hamilton James.

## Series editors

The following is a chronological list of series editors, the most senior member of the production team on *Natural World*[2] . Peter Jones, the first series editor, also acted in the same capacity on *The World About Us* from 1979-1983.

- Peter Jones (1983–1987)
- Andrew Neal (1987–1989)
- Mike Salisbury (1989–1993)
- John Sparks (1993–1997)
- Neil Nightingale (1997–2001)
- Mike Gunton (2001–2004)
- Tim Martin (2004–2011)
- Steve Greenwood (2011-present)

## Awards

*Natural World* filmmakers are regularly recognised by the television industry for the quality of their work. One of the earliest awards was the 1986 Prix Italia for Ecology for a special edition on soil erosion in the *Vanishing Earth* mini-series. In 1998, the Royal Television Society awarded the series the Best Documentary Strand Award.

In addition, the following individual films have recently won awards at the Missoula International Wildlife Film Festival, Jackson Hole Wildlife Film Festival, Wildscreen Festival, Grierson British Documentary Awards and RTS Awards:

- "Earth Pilgrim" (2008)
- "Lobo: The Wolf that Changed America" (2008)
- "Snow Leopard: Beyond the Myth" (2008)
- "Reindeer Girls" (2008)
- "White Falcon, White Wolf" (2008)
- "Battle to Save the Tiger" (2007)
- "Buddha, Bees and the Giant Hornet Queen" (2007)
- "Bonobo: Missing in Action" (2006)
- "The Orang-Utan King" (2005)
- "The Queen of Trees" (2005)[8]
- "Mississippi: Tales of the Last River Rat" (2004)
- "The Elephant, the Emperor and the Butterfly Tree" (2003)
- "Missing - Presumed Eaten" (2003)
- "Cats under Serengeti Stars" (2003)
- "My Halcyon River" (2002)
- "Danger in Tiger Paradise" (2002)
- "Warriors of the Monkey God" (2000)
- "Elephants of the Sand River" (1999)
- "Dolphins - The Wild Side" (1999)
- "The Dragons of Galapagos" (1998)
- "People of the Sea" (1997)
- "Lions - Pride in Peril" (1996)
- "The Call of Kakadu" (1995)

## DVD and video

Some episodes of *Natural World* were released on VHS (see Notable Episodes section for examples) but all are now out of print.

A Region 2 DVD, *The BBC Natural World Collection*, was released in 2007 and contains the following six episodes from the 2006 and 2007 seasons)[9] :

- "On the Trail of Tarka" 2006-10-25
- "The Bear Man of Kamchatka" 2006-11-08
- "Invasion of the Crocodiles" 2007-05-09
- "Eye for an Elephant" 2006-11-01
- "Buddha, Bees and the Giant Hornet Queen" 2007-04-25
- "Wye - Voices from the Valley" 2007-06-13

The following episodes were released on individual Region 2 DVDs in 2008:

- "Badgers - Secrets of the Sett" (BBCDVD2776)
- "Snow Leopard: Beyond the Myth" (BBCDVD2779)
- "Lobo - The Wolf That Changed America" (BBCDVD2780)
- "On The Trail Of Tarka" (BBCDVD2781)
- "The Bear Man of Kamchatka" (BBCDVD2782)
- "Earth Pilgrim: A Year on Dartmoor" (BBCDVD2790)

*BBC Natural World Collection* DVD cover

The UK releases of *Planet Earth* on HD DVD and Blu-ray includes the following 2 episodes of *Natural World*, both narrated by David Attenborough, in 1080i on a bonus disc:[10]

- "Desert Lions" 2007-05-30
- "Snow Leopard: Beyond the Myth" 2008-01-04

## References

[1] http://www.bbc.co.uk/programmes/b006qnnh
[2] "Natural World celebrates 25 years" (http://www.bbc.co.uk/pressoffice/pressreleases/stories/2008/10_october/23/natural.shtml). BBC Press Office. 2008-10-23. .
[3] "Oral History Chapter 8: The Arrival of Colour Television" (http://www.wildfilmhistory.org/oh/13/Barry+Paine.html). WildFilmHistory. 2001-01-31. .
[4] Attenborough, David (2002). *Life on Air: Memoirs of a Broadcaster*. BBC Books. p. 212. ISBN 0563487801.
[5] "Natural World celebrates 25th with three-year order" (http://www.broadcastnow.co.uk/news/2008/10/natural_world_wins_threeyear_order.html). Broadcast. 2008-10-21. .
[6] Bright, Michael (2007). *100 Years of Wildlife*. BBC Books. p. 46. ISBN 1846073219.
[7] Bouse, Derek (2000). *Wildlife Films*. University of Pennsylvania Press. pp. 14–15. ISBN 0812217284.
[8] "Treat for wildlife enthusiasts" (http://www.hindu.com/2006/09/08/stories/2006090805230200.htm). *The Hindu*. 2006-09-08. .
[9] "*The BBC Natural World Collection*" (http://www.bbcshop.com/invt/bbcdvd2363&bklist=icat,3,,281). www.bbcshop.com. .
[10] "Amazon.co.uk: Planet Earth : Complete BBC Series HD DVD" (http://www.amazon.co.uk/dp/B000SKLI2G). .

## External links

- *Natural World* (http://www.bbc.co.uk/programmes/b006qnnh) at BBC Programmes
- *Natural World* (http://www.imdb.com/title/tt0318912/) at the Internet Movie Database
- PBS website about *The Queen of Trees* episode (http://www.pbs.org/wnet/nature/queenoftrees/)

# Reverence

**Reverence** may refer to:

- Reverence (emotion) a subjective response to something excellent in a personal way
- Reverence (attitude), the acknowledgement of the legitimacy of the power of one's superior or superiors
- *Reverence* (Emperor album), a 1997 symphonic black metal album
- *Reverence* (Faithless album), a 1996 trip-hop album
- *Reverence* (The Tender Box album), a 2011 alternative rock album
- Reverence (music festival), an American electronic music festival
- Reverence (sculpture), an outdoor sculpture in Vermont
- "Reverence" (song), a 1992 alternative rock song

# Tree worship

**Tree worship** (dendrolatry) refers to the tendency of many societies throughout history to worship or otherwise mythologize trees. Trees have played an important role in many of the world's mythologies and religions, and have been given deep and sacred meanings throughout the ages. Human beings, observing the growth and death of trees, the elasticity of their branches, the sensitivity and the annual decay and revival of their foliage, see them as powerful symbols of growth, decay and resurrection. The most ancient cross-cultural symbolic representation of the universe's construction is the world tree.

The image of the Tree of life is also a favourite in many mythologies. Various forms of trees of life also appear in folklore, culture and fiction, often relating to immortality or fertility. These often hold cultural and religious significance to the peoples for whom they appear. For them, it may also strongly be connected with the motif of the world tree.

Other examples of trees featured in mythology are the Banyan and the Peepal (*Ficus religiosa*) trees in Hinduism, and the modern tradition of the Christmas Tree in Germanic mythology, the Tree of Knowledge of Judaism and Christianity, and the Bodhi tree in Buddhism. In folk religion and folklore, trees are often said to be the homes of tree spirits. Historical Druidism as well as Germanic paganism appear to have involved cultic practice in sacred groves, especially the oak. The term *druid* itself possibly derives from the Celtic word for oak.

Trees are a necessary attribute of the archetypical *locus amoenus* in all cultures. Already the Egyptian Book of the Dead mentions sycomores as part of the scenery where the soul of the deceased finds blissful repose.[1]

The evidence for tree-worship is almost unmanageably large, and since comparative studies do not as yet permit a concise and conclusive synopsis of the subject, this article will confine itself to some of the more prominent characteristics.

## Wishing trees

In almost every part of the world travelers have observed the custom of hanging objects upon trees in order to establish some sort of a relationship between themselves and the tree. Throughout Europe also, a mass of evidence has been collected testifying to the lengthy persistence of superstitious practices and beliefs concerning them. The trees are known as the scenes of pilgrimages, ritual ambulation, and the recital of (Christian) prayers. Wreaths, ribbons or rags are suspended to win favor for sick men or cattle, or merely for good luck. Popular belief associates the sites with healing, bewitching, or mere wishing; and though now perhaps the tree is the object only of some vague respect, there are abundant allusions to the earlier vitality of coherent and systematic cults. Decayed or fragmentary though the features may be in Europe.

Modern observers have found in other parts of the world more organic examples which enable us, not necessarily to reconstruct the fragments which have survived in the later religions and civilizations, but at least to understand their earlier significance. In India, for example, the Korwas hang rags on the trees which form the shrines of the village-gods. In Nebraska the object of the custom was to propitiate the supernatural beings and to procure good weather and hunting. In South America Darwin recorded a tree honored by numerous offerings (rags, meat, cigars, etc.); libations were made to it, and horses were sacrificed.[2] If, in this instance, the Gauchos regarded the tree, not as the embodiment or abode of Walleechu, but as the very god himself, this is a subtle but very important transference of thought, the failure to realize which has not been confined to those who have venerated trees.

## World tree

The tree, with its branches reaching up into the sky, and roots deep into the earth, can be seen to dwell in three worlds - a link between heaven, the earth, and the underworld, uniting above and below. It is also both a feminine symbol, bearing sustenance; and a masculine, phallic symbol - another union.

For this reason, many mythologies around the world have the concept of the World tree, a great tree that acts as an Axis mundi, supporting or holding up the cosmos, and providing a link between the heavens, earth and underworld. In European mythology the best known example is the tree Yggdrasil from Norse mythology.[3]

The world tree is also a central part of Mesoamerican mythologies, where it represents the four cardinal directions. The concept of the world tree is also closely linked to the motif of the Tree of life.

Yggdrasil, the World Ash (Norse)

## Popular stories

Numerous popular stories reflect a firmly rooted belief in an intimate connection between a human being and a tree, plant or flower. Sometimes a man's life depends upon the tree and suffers when it withers or is injured, and we encounter the idea of the external soul, already found in the Egyptian Tale of the Two Brothers of at least 3000 years ago. Here one of the brothers leaves his heart on the top of the flower of the acacia and falls dead when it is cut down. Sometimes, however, the tree is an index, a mysterious token which shows its sympathy with an absent hero by weakening or dying, as the man becomes ill or loses his life. These two features very easily combine, and they agree in representing to us mysterious sympathy between tree and human-life, which, as a matter of fact, frequently manifests itself in recorded beliefs and customs of historical times.

Thus, sometimes the new-born child is associated with a newly planted tree with which its life is supposed to be bound up; or, on ceremonial occasions (betrothal, marriage, ascent to the throne), a personal relationship of this kind is instituted by planting trees, upon the fortunes of which the career of the individual depends. Sometimes, moreover, boughs or plants are selected and the individual draws omens of life and death from the fate of his or her choice. Again, a man will put himself into relationship with a tree by depositing upon it something which has been in the closest contact with himself (hair, clothing, ext.). This is not so unusual as might appear; there are numerous examples of the conviction that a sympathetic relationship continues to subsist between things which have once been connected (e.g. a man and his hair), and this may be illustrated especially in magical practices upon material objects which are supposed to affect the former owner. We have to start then with the recognition that the notion of a real inter-connection between human life and trees has never presented any difficulty to primitive minds.

Often the tree is famous for oracles. Best known, perhaps, is the oak of Dodona tended by priests who slept on the ground. Forms of the tall oaks of the old Prussians were inhabited by gods who gave responses, and so numerous are the examples that the old Hebrew terebinth of the teacher, and the terebinth of the diviners may reasonably be placed in this category. Important sacred trees are also the object of pilgrimage, one of the most noteworthy being the branch of the Bo tree at Sri Lanka brought thither before the Christian era. The tree-spirits will hold sway over the surrounding forest or district, and the animals in the locality are often sacred and must not be harmed.

## Disease and demons

The custom of transferring disease or sickness from men to trees is well known. Sometimes the hair, nails, clothing, ext., of a sickly person are fixed to a tree, or they are forcibly inserted in a hole in the trunk, or the tree is split and the patient passes through the aperture. Where the tree has been thus injured, its recovery and that of the patient are often associated. Different explanations may be found of such customs which naturally take rather different forms among peoples in different grades.

In India, for example, when the patient is supposed to be tormented by a demon, ceremonies are performed to provide it with a tree where it will dwell peacefully without molesting the patient so long as the tree is left unharmed. Such ideas do not enter, of course, when the rite merely removes the illness and selfishly endangers the health of those who may approach the tree. Again, sometimes it is clearly felt that the main personality has been mystically united with some healthy and sturdy tree, and in this case we may often presume that such trees already possessed some peculiar reputation. The custom finds an analogy when hair, nail-clippings, ext., are hung upon a tree for safety sake lest they fall into the hands of an enemy who might injure the owner by means of them.

Among the Arabs the sacred trees are haunted by angels or by jinn; sacrifices are made, and the sick who sleep beneath them receive prescriptions in their dreams. Here, as frequently elsewhere, it is dangerous to pull a bough. This dread of damaging special trees is familiar: Cato instructed the woodman to sacrifice to the male or female deity before thinning a grove, while in the Homeric poem to Aphrodite the tree nymph is wounded when the tree is injured, and dies when the trunk falls.

Early Buddhism decided that trees had neither mind nor feeling and might lawfully be cut; but it recognized that certain spirits might reside in them, and this the modern natives of India firmly believe. Propitiation is made before the sacrilegious axe is laid to the holy trees; loss of life or of wealth and the failure of rain are feared should they be wantonly cut; there are even trees which it is dangerous to climb. The Talein of Burma prays to the tree before he cuts it down, and the African woodman will place a fresh sprig upon the tree.

## Sacred trees

Stefan of Perm takes an axe to a birch hung with pelts and cloths that is sacred to the Komi of Great Perm

Trees were often regarded as sacred in the ancient world, throughout Europe and Asia[4] . Christianity and Islam treated the worship of trees as idolatry and this led to their destruction in Europe and most of West Asia. In the manuscript illumination *(illustration)* Saint Stephan of Perm cuts down a birch sacred to the Komi people as part of his proselytizing among them in the years after 1383. His profanation of their shrines and cult images incurred their hostility.[5]

Sacred trees remain common in India. They are found in villages, in the countryside and the heart of some temples (e.g. Jain temples).

In Sri Lanka, Buddhists worship the Bodhi Tree. It is said to have protected the Buddha when he was meditating to attain enlightenment.

The Glastonbury Thorn in Glastonbury, England is a small Common Hawthorn tree regarded as sacred by many Christians. It is said to have sprouted miraculously from the staff of the early Christian figure Joseph of Arimathea. Of further religious significance and indeed scientific interest, the tree displays a rare phenomenon for its species, blooming not once but twice per year. The second bloom occurs around the holiday of Christmas.

## Sacred groves

Many of the world's ancient belief systems also include the belief of sacred groves, where trees are revered and respected and there are priests and priestesses attending to them who also serve as guardians, preventing those who wish to tear down the trees by means of ancient magic and elaborate protection rituals.

From ancient Norse and Celtic mythologies, to the Nigerian and Indian cosmological thoughts, extending far east in the ancient Shinto faith of Japan and the peculiar habits of the 19 tribes of the forest peoples of Malaysia, sacred groves provide relief and shelter from the mundane aspects of life and are considered living temples, albeit absent of stone walls or ornate stone monuments. A place of meeting where ancient rituals are performed, it is also a place of refuge for many in times of danger. For those who were fated not find peace in this life, it is considered as the final resting place where the soul finds eternal peace as it reunites with the creator.

## In literature

In literature, a mythology was notably developed by J. R. R. Tolkien, his Two Trees of Valinor playing a central role in his mythopoeic cosmogony. Tolkien's 1964 *Tree and Leaf* combines the allegorical tale *Leaf by Niggle* and his essay *On Fairy-Stories*. William Butler Yeats describes a "holy tree" in his poem *The Two Trees* (1893). Also in the Lord of the Rings, the White Tree stands symbol to Minas Tirith and Gondor. In Laurie Halse Anderson's novel *Speak*, the main character Melinda Sordino is assigned by her art teacher to make a tree "into a piece of art" (p. 12 Mr. Freeman talking to art class). The tree Melinda tries to draw and sculpt but keeps failing becomes a symbol of Melinda's internal battle of wanting to speak up about the rape that was inflicted upon her but continues to remain silent. When Melinda succeeds in creating a tree (and gets an A+), it symbolizes Melinda, finally speaking out against the injustice.

## See also

- Axis mundi
- Celtic tree worship
- Christmas Tree
- Five Trees
- Gerichtslinde
- Irminsul
- Mesoamerican world tree
- Nature worship
- New Year Tree
- Sacred garden
- Sacred grove
- Sacred herbs
- Sephirot (Kabbalah)
- Sidrat al-Muntaha
- Talking trees
- Tree of life
- Tree of life (biblical)
- Tree of Life (Kabbalah)
- Tree of Knowledge of Good and Evil
- Thor's oak
- Wish Tree
- World tree

## References

[1] Gollwitzer 1984:13.
[2] "The Voyage of the Beagle" (http://www.bartleby.com/29/4.html), Chapter IV
[3] Mountfort 2003:41, 279.
[4] Gorshunova. Olga V. (2008). *Svjashennye derevja Khodzhi Barora...*, ( *Sacred Trees of Khodzhi Baror: Phytolatry and the Cult of Female Deity in Central Asia*) in Etnoragraficheskoe Obozrenie, n° 1, pp. 71-82. ISSN 0869-5415. (Russian)
[5] Forsyth 1992:5f.

## Other sources

- Becker, Lore (2002). *Die Mythologie der Bäume*, Papyrus 1-2 (http://www.papyrus-magazin.de/archiv/2002_2003/november/11_12_2002_mythologie1.html)
- Brosse, Jaques (1989). *Mythologie des arbres*, ISBN 978-2228887113.
- Forsyth, James (1992). *A History of the Peoples of Siberia: Russia's North Asian Colony 1581-1990* (http://books.google.com/books?id=nzhq85nPrdsC&printsec=frontcover&source=gbs_ge_summary_r&cad=0#v=onepage&q&f=false). Cambridge University Press. ISBN 0-521-47771-9
- Forlong James Rivers of Life London & Edinburgh 1883 vol I chapter 2- Tree Worship
- Gollwitzer, Gerda (1984). Botschaft der Bäume, DuMont Buchverlag Köln
- Hageneder, Fred (2005). *The Meaning of Trees: Botany, History, Healing, Lore* (http://books.google.com/books?id=M_a1mYTeQXsC&printsec=frontcover&source=gbs_ge_summary_r&cad=0#v=onepage&q&f=false). Chronicle Books. ISBN 0-8118-4823-X.
- Malla, Bansi Lal (2000). *Trees in Indian Art, Mythology, and Folklore*, ISBN 8173051798.
- Mountfort, Paul Rhys (2003). *Nordic Runes: Understanding, Casting, and Interpreting the Ancient Viking Oracle* (http://books.google.co.in/books?id=_3B7EmvAqngC). Inner Traditions / Bear & Company.

ISBN 0-89281-093-9.

- Porteous, Alexander (2002). *The Forest in Folklore and Mythology* (http://books.google.com/books?id=PyioYIjLWy8C&printsec=frontcover&source=gbs_ge_summary_r&cad=0#v=onepage&q&f=false). Courier Dover Publications. ISBN 0-486-42010-8.

This article incorporates text from a publication now in the public domain: Chisholm, Hugh, ed (1911). *Encyclopædia Britannica* (11th ed.). Cambridge University Press.

# Animal worship

**Animal worship** (or **zoolatry**) refers to religious rituals involving animals, especially in pre-modern societies, such as the glorification of **animal deities**, or **animal sacrifice**.

The origins of animal worship have been the subject of many theories. The classical author Diodorus explained the origin of animal-worship by recalling the myth in which the gods, supposedly threatened by giants, hid under the guise of animals. The people then naturally began to worship the animals that their gods had disguised themselves as and continued this act even after the gods returned to their normal state (Lubbock, 2005, p. 252). In 1906, Weissenborn suggested that animal worship resulted from man's natural curiosity. Primitive man would observe an animal that had a unique trait and the inexplicability of this trait would appeal to man's curiosity (Weissenborn, 1906b, p. 282). Wonder resulted from primitive man's observations of this distinctive trait and this wonder eventually induced adoration. Thus, primitive man worshipped animals that had inimitable traits (Weissenborn, 1906b, p. 282). Lubbock put forward a more recent view. Lubbock proposed that animal-worship originated from family names. In societies, families would name themselves and their children after certain animals and eventually came to hold that animal above other animals. Eventually, these opinions turned into deep respect and evolved into fully developed worship of the family animal (Lubbock, 2005, p. 253). The belief that an animal is sacred frequently results in dietary laws prohibiting their consumption. As well as holding certain animals to be sacred, religions have also adopted the opposite attitude, that certain animals are unclean.

The idea that divinity embodies itself in animals, such as a deity incarnate, and then lives on earth among human beings is disregarded by Abrahamic religions (Morris, 2000, p. 26). In churches such as Independent Assemblies of God and Pentecostal, animals have very little religious significance (Schoffeleers, 1985; Peltzer, 1987; Qtd. in Morris, 2000, p. 25). Animals have become less and less important and symbolic in cult rituals and religion, especially among African cultures, as Christianity and Islamic religions have spread. (Morris, 2000, p. 24).

The Egyptian pantheon was especially fond of zoomorphism, with many animals sacred to particular deities—cats to Bastet, ibises and baboons to Thoth, crocodiles to Sebek and Ra, fish to Set, mongoose, shrew and birds to Horus, dogs and jackals to Anubis, serpents and eels to Atum, beetles to Khepera, bulls to Apis. Animals were often mummified as a result of these beliefs.

## Classification of animal cults

When a god is respected or worshipped by means of a representative animal, an animal cult is formed (Teeter et al., 2002, p. 355). Such cults may result from early man's awareness of his essential equality with animals (Raglan, 1935, p. 331). Given this fundamental equality, it was just as reasonable for them to represent their gods in animal form (Gilbert, Qtd. in Raglan, 1935, p. 331).

Animal cults may be classified in two ways:

- according to their outward form;
- according to their inward meaning, which may of course undergo transformations.

## By outward form

There are two broad divisions:

1. all animals of a given species are sacred, perhaps owing to the impossibility of distinguishing the sacred few from the profane crowd; (Sacred-profane dichotomy)
2. one or a fixed number of a species are sacred. It is probable that the first of these forms is the primary one and the second in most cases a development from it due to
   1. the influence of other individual cults,
   2. anthropomorphic tendencies,
   3. the influence of chieftainship, hereditary and otherwise,
   4. annual sacrifice of the sacred animal and mystical ideas connected therewith,
   5. syncretism, due either to unity of function or to a philosophical unification,
   6. the desire to do honour to the species in the person of one of its members, and possibly other less easily traceable causes.

## By inward meaning

Treating cults according to their meaning, which is not necessarily identical with the cause which first led to the deification of the animal in question, we can classify them under ten specific heads:

- **Pastoral cults**: The pastoral type falls into two sub-types, in which the species is spared and sometimes receives special honour at intervals in the person of an individual. (See *Cattle*, *Buffalo*, below.)
- **Hunting cults**: In hunting cults the species is habitually killed, but occasionally honoured in the person of a single individual, or each slaughtered animal receives divine honours. (See *Bear*, below.)
- **Dangerous or noxious animals**: The cult of dangerous animals is due to the fear that the soul of the slain beast may take vengeance on the hunter, to a desire to placate the rest of the species. (See *Leopard*, below.)
- **Animals regarded as human souls or their embodiment**:Animals are frequently regarded as the abode, temporary or permanent, of the souls of the dead, sometimes as the actual souls of the dead. Respect for them is due to two main reasons: the kinsmen of the dead desire to preserve the goodwill of their dead relatives and they wish at the same time to secure that their kinsmen are not molested and caused to undergo unnecessary suffering. (See *Elephant*, below.)
- **Totemistic cults**: One of the most widely found modes of showing respect to animals is known as totemism (see totem), where a particular animal is seen as sacred to an individual or group, but except in decadent forms there is but little positive worship. In Central Australia, however, the rites of the Wollunqua totem group are directed towards placating this mythical animal, and cannot be termed anything but religious ceremonies. In secret societies we find bodies of men grouped together with a single tutelary animal; the individual, in the same way, acquires the nagual or individual totem, sometimes by ceremonies of the nature of the bloodbond. While an individual's or group's totem is sacred to them, it would not be sacred to another individual or group. By contrast, sacred animals in other religions are thus seen as sacred to the gods, and thus to all humanity, rather than particular individuals or tribes, even though only those sharing this belief will acknowledge their sacredness.
- **Cults of nature and vegetation spirits**: Spirits of the landscape such as vegetation, rain and the earth in many parts of the world, such as Europe, Chine and Mesoamerica are conceived in animal form. (See *Goat*, *Serpent* below.)
- **Cults of ominous animals**:The ominous (in the sense of being an omen) animal or bird may develop into a deity. (See *Hawk*, below.)
- **Cults of animals associated with zoomorphic deities**: It is commonly assumed that the animals associated with certain deities are sacred because the god was originally zoomorphic or theriomorphic (shapeshifting) ; this is doubtless the case in certain instances; but Apollo Smintheus, Dionysus Bassareus and other examples seem to show that the god may have been appealed to for help and thus become associated with the animals from whom

he protected the crops, and so on.

- **Cults of animals used in magic**: The use of animals in magic may sometimes give rise to a kind of respect for them, but this is of a negative nature. See, however, articles by Preuss in Globus, vol. lxvii., in which he maintains that animals of magical influence are elevated into divinities.

## Hunting cults

### Bear

The bear enjoys a large measure of respect from all cultures that come in contact with it, which shows itself in apologies and in festivals in its honour. Bear worship is found in a number of religions, including Finnish paganism. The most notable ceremonies involving bears are found in East Asia.

There is a festival among the Nivkhs that takes the form of a celebration in honour of a recently dead kinsman, to whom the spirit of the bear is sent. There have been some attempts to revive the practice.

There is a good deal of evidence to connect the Greek goddess Artemis with a cult of the bear. Girls danced as "bears" in her honour, and might not marry before undergoing this ceremony. According to mythology, the goddess once transformed a nymph into a bear and then into the constellation Ursa Major.

The bear is traditionally associated with Bern, Switzerland. It is believed that the city's name derives from the Germanic word for "bears" (*Bären* in German) and a bear is featured on the city's flag and coat of arm. In 1832 a statue of the Celtic bear goddess Artio was dug up there.

The existence of an ancient bear cult among Neanderthals in the Middle Paleolithic period has been a topic of discussion spurred by archaeological findings (Wunn, 2000, p. 434-435). Ancient bear bones have been discovered in several different caves and their peculiar arrangement are believed by some archaeologists to be evidence of a bear cult during the Paleolithic era. (Wunn, 2000, p. 435). However, others argue that the placement of these remains, whether it appears to be an identified pattern or not, is due to natural causes such as wind, sediment, or water (Wunn, 2000, p. 437-438).

The Ainu *Iomante* ceremony (bear sending). Japanese scroll painting, circa 1870.

The Ainu people, who live on select islands in the Japanese archipelago, call the bear "kamui" in their language, which translates to mean god. While many other animals are considered to be gods in the Ainu culture, the bear is the head of the gods (Kindaichi, 1949, p. 345). For the Ainu, when the gods visit the world of man, they don fur and claws and take on the physical appearance of an animal. Usually, however, when the term "kamui" is used, it essentially means a bear (Kindaichi, 1949, p. 345). The Ainu people willingly and thankfully ate the bear as they believed that the disguise (the flesh and fur) of any god was a gift to the home that the god chose to visit (Kindaichi, 1949, p. 348).

### Whale

In middle and South Vietnam coastal area, whale is considered sacred creature, who bring lucks and prosperity to fishermen. They are respectfully addressed as "the Lord". Coastal villages often hold funerals for beached whales, their remains are buried and skeletons are collected to be revered.

The first emperor of the last dynasty of Vietnam is believed to be rescued by a whale when his ship was capsized in a storm. After unifying the country, he sainted the whale who came to his rescue, which is known today as "Đông Hải phúc thần" (the deity of Fortune in the East sea).

A prevalent whale cult in Japan occurs around the coastal area. There are cemeteries with memorial stones dedicated to the whales which were hunted and killed to feed the people (Naumann, 1974, p. 4). Buddhist epitaphs mark these stones which implore that Buddha be reborn as a whale (Naumann, 1974, p. 4). Along with these memorials, there is evidence that whale embryos, found in a deceased mother's womb, were extracted and buried with the same respect as a human being (Naumann, 1974, p. 5). For certain shrines, the bones of a perished whale were also deposited in the area (Naumann, 1974, p. 5).

In Alaska, there were certain tribes that had ceremonial tributes to pieces of a whale after it was captured in a hunt (Lantis 1938, p. 445). Some tribes brought the hump, the fins, or the nose of the whale into their camps or the whaler's house. These parts were meant to represent the entirety of the whale and were honored as such during the festival (Lantis 1938, p. 445). The bones of a whale, however, were also given ritual treatment. The Alaskan tribes that participated in such acts believed that their rituals protected the whale's soul from injury and it could then be free to return to the sea (Lantis 1938, p. 445).

In the book Moby Dick can be found a longer description of the symbolism of the whale.

## Domesticated mammals

### Cattle and buffalo

Many religions have considered cattle to be sacred, most famously Hinduism from Nepal and India, but also Zoroastrianism, and ancient Greek and Egyptian religion. Cattle and buffalo are respected by many pastoral peoples that rely on the animals for sustenance and the killing of an ox is a sacrificial function.

In Hinduism, the cow is a symbol of wealth, strength, abundance, selfless giving and a full Earthly life

The Toda of southern India abstain from the flesh of their domestic animal, the buffalo. However, once a year they sacrifice a bull calf, which is eaten in the forest by the adult males. The buffalo plays an important part in many Toda rituals. These buffalo are currently endangered.

Conspicuous among Egyptian animal cults was that of the bull, Apis. It was distinguished by certain marks, and when the old Apis died a new one was sought. The finder was rewarded, and the bull underwent four months' education at Nilopolis. Its birthday was celebrated once a year when oxen, which had to be pure white, were sacrificed to it. Women were forbidden to approach it when once its education was finished. Oracles were obtained from it in various ways. After death it was mummified and buried in a rock-tomb. Less widespread was the cult of the Mnevis, also consecrated to Osiris.

Similar observances are found in our own day on the Upper Nile. The Nuba and Nuer revere cattle. The Angoni of Central Africa and the Sakalava of Madagascar keep sacred bulls. In India respect for the cow is widespread, but is of post-Vedic origin; there is little actual worship, but the products of the cow are important in magic.

While there are several animals that are worshipped in India, the supreme position is held by the cow (Margul, 1968, p. 63). The humped zebu, a breed of cow, is central to the religion of Hinduism (Margul, 1968, p. 63). Mythological legends have supported the sanctity of the zebu throughout India (Margul, 1968, p. 64). Such myths have included the creation of a divine cow mother and a cow heaven by the God, Brahma and Prithu, the sovereign of the universe, created the earth's vegetation, edible fruits and vegetables, disguised as a cow (Margul, 1968, p. 64).

According to Tadeusz Margul, observations of the Hindu religion and the cow has led to a misunderstanding that Hindi have a servile relationship with the zebu, giving prayers and offerings to it daily. Typically, however, only during the Cow Holiday, an annual event, is the cow the recipient of such practices (Margul, 1968, p. 65). Margul suggests that sanctity of the cow is based on four foundations: abstaining from cow slaughter, abstaining from beef consumption, control of breeding and ownership, and belief in purification qualities of cow products (milk, curd, ghee, dung, and urine) (Margul, 1968, p. 65-66).

## Sheep

Only in Africa do we find a sheep-god proper. Amun, the god of Thebes, Egypt, was represented as ram-headed. His worshippers held the ram to be sacred, however, it was sacrificed once a year. Its fleece formed the clothing of the idol. Another Egyptian ram-headed god was *Banebdjed*, a form of Osiris.

## Goat

Pavement mosaic with the head of Pan. Roman artwork, Antonine period, 138–192 CE.

Silenus, the Satyrs and the Fauns were either capriform or had some part of their bodies shaped like that of a goat. In northern Europe the wood spirit, Leszi, is believed to have a goat's horns, ears and legs. In Africa the Bijago people are said to have a goat as their principal divinity. A deity known as the Goat of Mendes is associated with the pentagram.

In Greece, Italy, and Egypt, the goat was worshipped in both goat form and phallic form (Neave 1988, p. 8). This type of worship has sometimes been said to have originated from the goat's increased sex drive. One male goat was capable of fertilizing 150 females (Neave 1988, p. 8). The Greek god Pan was depicted as having goat characteristics, such as hooves, horns, and a beard. Along with Pan, the goat was closely related to Dionysus during the Roman era (Neave 1988, p. 8). To honor Dionysus, Romans would tear apart a goat and eat it alive. The goat was commonly associated with dark arts and the devil. This association was amplified in Europe during the Middle Ages (Neave 1988, p. 8).

Excavations in Central Asia have revealed ancient ritual goat-burial that show a religious significance of the goat predominantly in the Afghanistan area (Sidky 1990, p. 286). These findings have been used as evidence for a goat-cult of Asia originating either in the Neolithic or Bronze Ages (Sidky 1990, p. 286). An example of a goat-burial finding was discovered on the Oxus River. Along these banks, a Neanderthal grave was excavated surrounded by several pairs of goat horns, suggesting significance of goats in Central Asian religion (Sidky 1990, p. 286).

## Dog

Dogs have a major religious significance among the Hindus in Nepal and some parts of India. The dogs are worshipped as a part of a five-day Tihar festival that falls roughly in November every year. In Hinduism, it is believed that the dog is a messenger of *Yama*, the god of death, and dogs guard the doors of Heaven. Socially, they are believed to be the protectors of our homes and lives. So, in order to please the dogs they are going to meet at Heaven's doors after death, so they would be allowed in Heaven, people mark the 14th day of the lunar cycle in November as **Kukur-tihar**, as known in Nepali language for the dog's day. This is a day when the dog is worshipped by applying *tika* (the holy vermilion dot), incense sticks and garlanded generally with marigold flower.

A dog after being decorated in Kukur tihar festival in Nepal

Actual dog worship is uncommon. The Nosarii of western Asia are said to worship a dog. The Karang of Java had a cult of the red dog, each family keeping one in the house. According to one authority the dogs are images of wood which are worshipped after the death of a member of the family and burnt after a thousand days. In Nepal it is said that dogs are worshipped at the festival called Khicha Puja. Among the Harranians dogs were sacred, but this was rather as brothers of the mystae.

## Horse

Horse worship has been practiced by a number of Indo-European and Turkic peoples. There is some reason to believe that Poseidon, like other water gods, was originally conceived under the form of a horse. In the cave of Phigalia Demeter was, according to popular tradition, represented with the head and mane of a horse, possibly a relic of the time when a non-specialized corn-spirit bore this form. Her priests were called Poloi (Greek for "colts") in Laconia. The mule and the horse are sacred to the Roman god Consus. In Gaul(France) we find a horse-goddess, Epona. There are also traces of a horse-god, Rudiobus. Hayagriva is a horse-headed deity that appears in both Hinduism and Buddhism. The Gonds in India worship a horse-god, Koda Pen, in the form of a shapeless stone, but it is not clear that the horse is regarded as divine. The horse or mare is a common form of the corn-spirit in Europe.

Among the Balkan culture, swaddling an unmarried person in a horse-girth is a typical ritual. It is thought that the sexual potency of the horse is passed to the individual wrapped in its girth (Vukanović 1980, p. 112). Along with the Balkan swaddling, Virgil's Aeneid bases the founding of the great city of Carthage upon a horse (Qtd. in Brown 1950, p. 32). When the Phoenicians dug up a horse head from the ground they decided to build their city (Carthage) upon that spot because the horse was a sign of success (Qtd. in Brown 1950, p. 32). Thus, Brown argued that the horse was sacred to the Phoenician people (Brown 1950, p. 32).

### Elephant

In Thailand it is believed that a white elephant may contain the soul of a dead person, perhaps a Buddha. When one is taken the capturer is rewarded and the animal brought to the king to be kept ever afterwards. It cannot be bought or sold. It is baptized and fêted and mourned for like a human being at its death. In some parts of Indo-China the belief is that the soul of the elephant may injure people after death; it is therefore fêted by a whole village. In Cambodia it is held to bring luck to the kingdom. The cult of the white elephant is also found at Ennarea in southern Ethiopia. In India, the popular Hindu god Ganesha has the head of an elephant and a torso of a human.

A statue of Ganesha - the elephant headed Hindu god of wisdom and obstacle removal

In Surat, unmarried Anāvil girls participate in a holiday referred to as Alunām (Naik, 1958, p. 393). This holiday is to honor the goddess Pārvatī. During this celebration, a clay elephant is prepared (most likely to celebrate Pārvatī's creation of Ganesha from a paste of either tumeric or sandalwood). Every day, the unmarried women worship this elephant by dancing, singing songs, and abstaining from eating salt. On the final day of Alunām, the clay elephant is immersed in some body of water (Naik, 1958, p. 393).

Certain cultures also used elephant figurines to display the animal's importance. There was evidence of an ancient elephant cult in Sumatra (Schnitger, 1938, p. 41). Stone elephant figurines were built as "seats of the souls" in the Sumatran culture (Schnitger, 1938, p. 41). In North Borneo, however, wooden elephant figurines were placed on the top of a bamboo pole. This bamboo pole was only erected after the tribe chief had collected a certain number of human heads (Schnitger, 1938, p. 41).

## Wild mammals

In North America the Algonquian tribes had as their chief deity a "mighty great hare" to whom they went at death. According to one account he lived in the east, according to another in the north. In his anthropomorphized form he was known as Menabosho or Michabo.

The deer is important in the mythology of many peoples. To the Greeks it was sacred to the goddess Artemis, while in Hinduism it is linked to the goddess Saraswati. The deer also held spiritual significance to the pastoralist cultures of the Eurasian Steppe. The golden stag figurine found in the Pazyryk burials is one of the most famous pieces of Scythian art.

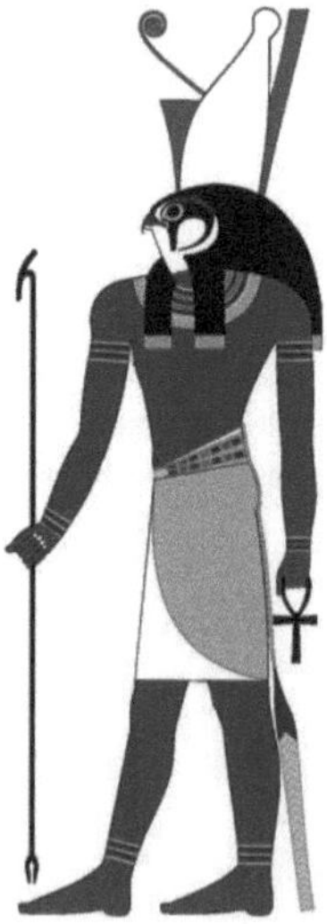

Horus

### Tiger

In many parts of Vietnam, tiger is a revered creature. In each village, there might be a tiger temple. This worshiping practice might have stem from the fear of tigers used to raid human settlements in the ancient time. Tigers are admired for their great strength, ferocity and grace. Tiger is also considered a guardian deity. Tiger statutes are usually seen at the entrance of temples and palaces, keeping evil spirits from entering those places.

## Wolf

The cult of the wolf is frequently found among the tutelary deities of North American dancing or secret societies. The Tlingit had a god, Khanukh, whose name means "wolf," and worshipped a wolf-headed image.

In Altaic mythology of the Turkic and Mongolian peoples, the wolf is a revered animal. The shamanic Turkic peoples even believed they were descendants of wolves in Turkic legends. The legend of Asena is an old Turkic myth that tells of how the Turkic people were created. In Northern China a small Turkic village was raided by Chinese soldiers, but one small baby was left behind. An old she-wolf with a sky-blue mane named Asena found the baby and nursed him, then the she-wolf gave birth to half wolf, half human cubs therefore the Turkic people were born. Also in Turkic mythology it is believed that a gray wolf showed the Turks the way out of their legendary homeland *Ergenekon*, which allowed them to spread and conquer their neighbours.

Both Zeus and Apollo were associated with the wolf by the Greeks, but it is not clear that this implies a previous cult of the wolf.

In Western culture, the most obvious example is the symbolism of Rome's foundation, and the use of wolves in totemic imagery.

## Cats

The cult of the leopard is widely found in West Africa. Among the Ashanti people a man who kills one is liable to be put to death; no leopard skin may be exposed to view, but a stuffed leopard is worshiped. On the Gold Coast a leopard hunter who has killed his victim is carried round the town behind the body of the leopard; he may not speak, must besmear himself so as to look like a leopard and imitate its movements. In Loango a prince's cap is put upon the head of a dead leopard, and dances are held in its honour.

During the Egyptian Twenty-sixth Dynasty people began mummifying particular animal species as offerings to the god whom the species represented. Millions of mummified cats, birds, and other creatures were buried at temples honoring Egyptian deities.[1] [2] Worshippers paid the priests of a particular deity to obtain and mummify an animal and the mummy was placed in a cemetery near the god's cult center. The lion was associated with the Egyptian deities Horus, Nefertum, Ra and Sekhmet. There was a lion-god at Baalbek . The pre-Islamic Arabs had a lion-god, Yaghuth. In modern Africa we find a lion-idol among the Balonda. The lion was also sacred to Hebat, the mother goddess of the Hurrians.

In Judaism the patriarch Jacob refers to his son Judah as a Gur Aryeh הָדוּהְי הֵיְרַא רוּג, a "Young Lion" (Genesis 49:9) when blessing him. Thus the Lion of Judah started to be reverenced in some others abrahamic cults, symbolising their prophets, as such as Jesus and Haile Selassie I, the *ras* Tafari.

In Mesoamerica the jaguar was revered as a symbol of fertility and warriorship among the Aztec, Maya and Olmec, and had an important role in shamanism.

Of great importance in Chinese myth and culture, the tiger is one of the 12 Chinese zodiac animals. Also in various Chinese art and martial art, the tiger is depicted as an earth symbol and equal rival of the Chinese dragon- the two representing matter and spirit respectively. The White Tiger (Chinese: ; pinyin: *Bái Hǔ*) is one of the Four Symbols of the Chinese constellations. It is sometimes called the White Tiger of the West (), and it represents the west and the autumn season.[3]

The tiger replaces the lion as King of the Beasts in cultures of eastern Asia,[4] representing royalty, fearlessness and wrath.[3]

Some cultures that celebrated tiger worship are still represented contemporarily. In the suburbs of Kunming, China, there is a tourist attraction where the tiger worship of the Yi is displayed for visitors. This attraction called the Solar Calendar Square is complete with a growling tiger statue, measuring to be five meters high (Harrell & Yongxiang 2003, p. 380). In Chuxiong of China, a similar attraction exists. A tiger totem is presented for tourists; the totem portrays the Yi belief of the tiger setting the entire world in motion. A tiger dance of the Shuangbai County is also performed at such places explaining the history of the Yi and their worship of tigers (Harrell & Yongxiang 2003, p. 380).

Along with these tourist attractions that display historical practices of the Yi, there is also additional evidence for tiger worship. Tigers were found depicted on small stones. These stones were pierced and worn as amulets, suggesting that the tiger had a certain power of protection for its wearer (Waterbury 1952, p. 76). The Queen Mother deity of the west, Hsi Wang Mu, sometimes possessed a tail of a tiger in her depictions and, like the tiger, was associated with the mountains (Waterbury 1952, p. 76). The tiger was also a deity for both the Tungus and the Black Pottery people (Waterbury 1952, p. 80).

The tiger is associated with the Hindu deities Shiva and Durga. In Pokhara, Nepal the tiger festival is known as Bagh Jatra. Celebrants dance disguised as tigers and "hunted". The Warali tribe of Maharashtra, India worship Waghia the lord of tigers in the form of a shapeless stone. In Hanoi and Manchuria tiger-gods are also found.

### Monkey

In India the monkey-servant of God, Hanuman, is a prominent figure.He is a deity born as the reincarnation of Lord Shiva, the god of destruction. In orthodox villages monkeys are safe from harm. Monkeys are said to be worshipped in Togo. At Porto Novo, in French West Africa, twins have tutelary spirits in the shape of small monkeys.

## Birds

### Crow/Raven

The Raven is the chief deity of the Tlingit people of Alaska. All over that region it is the chief figure in a group of myths, fulfilling the office of a culture hero who brings the light, gives fire to mankind, and so on. A raven story from the Puget Sound region describes the "Raven" as having originally lived in the land of spirits (literally *bird land*) that existed before the world of humans. One day the Raven became so bored with *bird land* that he flew away, carrying a stone in his beak. When the Raven became tired of carrying the stone and dropped it, the stone fell into the ocean and expanded until it formed the firmament on which humans now live.

In the creator role, and in the Raven's role as the totem and ancestor of one of the four northwest clan houses, the Raven is often addressed as *Grandfather Raven*. It is not clear whether this form of address is intended to refer to a creator Raven who is different from the trickster Raven, or if it is just a vain attempt to encourage the trickster spirit to act respectably.

Together with the eagle-hawk the crow plays a great part in the mythology of southeastern Australia. Ravens also play a part in some European mythologies, such as in the Celtic and Germanic Religions, where they were connected to Bran and the Morrigan in the former and Woden in the latter.

### Hawk

North Borneo treated the hawk as a god, but it was technically the messenger of the people's Supreme God (Waterbury 1952, p. 62). There were rituals that involved the hawk when the natives wished to make decisions about certain events, such as journeys from home, major agricultural work, and war (Waterbury 1952, p. 62). In North Borneo we seem to see the evolution of a god in the three stages of the cult of the hawk among the Kenyahs, the Kayans and the sea Dyaks. The Kenyahs will not kill it, address to it thanks for assistance, and formally consult it before leaving home on an expedition. It seems, however, to be regarded as the messenger of the supreme god Balli Penyalong. The Kayans have a hawk-god, Laki Neho, but seem to regard the hawk as the servant of the chief god, Laki Tenangan. Singalang Burong, the hawk-god of the Dyaks, is completely anthropomorphized. He is god of omens and ruler of the omen birds, but the hawk is not his messenger. For he never leaves his house. Stories are, however, told of his attending feasts in human form and flying away in hawk form when all was over.

According to Florance Waterbury, hawk worship was universal (Waterbury 1952, p. 26). This particular bird was "a heavenly deity; its wings were the sky, the sun and moon were its eyes" (Waterbury 1952, p. 26). The hawk is commonly associated with the Egyptian god Horus. The souls of former pharaohs were the followers of Horus and therefore, the hawk (Waterbury 1952, p. 26). Horus was depicted by the Egyptians as a human body with a hawk head after the Fourth and Fifth Dynasty, but before that he was represented as a hawk (Waterbury 1952, p. 27).

Egypt was not the only location of hawk worshippers. There were several other cultures which held the hawk in high regard. The hawk was a deity on the island of Hawaii and symbolized swift justice (Waterbury 1952, p. 62). Along with the lone island from the Hawaiian archipelago, the Fiji islands also had some tribes who worshipped a hawk god (Waterbury 1952, p. 62).

### Frigate Bird

On Easter Island until the 1860s there was a Tangata manu (Bird man) cult which has left us Paintings and Petroglyphs of Birdmen (half men half Frigate birds). The cult involved an annual race to collect the first Sooty Tern egg of the season from the islet of Moto Iti and take it to Orongo.

The Frigate Bird Cult is thought to have originated in the Solomon Islands before immigrating to Easter Island where it became obsolete (Balfour 1917, p. 374). The Frigate-Bird was a representation of the god Make-make, the god of the seabird's egg on Easter Island (Balfour 1917, p. 374).

## Other non-mammals

### Serpent

The worship of the serpent is found in many parts of the Old World, and in the Americas.

In India Snake worship refers to the high status of snakes in Hindu mythology. Over a large part of India there are carved representations of cobras (nagas) or stones as substitutes. To these human food and flowers are offered and lights are burned before the shrines. Among the Dravidians a cobra which is accidentally killed is burned like a human being; no one would kill one intentionally. The serpent-god's image is carried in an annual procession by a celibate priestess.

The altar where Serpent deities are worshipped in a temple in Belur, Karnataka, India

At one time there were many prevalent different renditions of the serpent cult located in India. In Northern India, a masculine version of the serpent named Nagaraja and known as the "king of the serpents" was worshipped. Instead of the "king of the serpents," actual live snakes were worshipped in South India (Bhattacharyya 1965, p. 1). The Manasa-cult in Bengal, India, however, was dedicated to the anthropomorphic serpent goddess, Manasa (Bhattacharyya 1965, p. 1).

In Africa the chief centre of serpent worship was Dahomey. but the cult of the python seems to have been of exotic origin, dating back to the first quarter of the 17th century. By the conquest of Whydah the Dahomeyans were brought in contact with a people of serpent worshippers, and ended by adopting from them the beliefs which they at first despised. At Whydah, the chief centre, there is a serpent temple, tenanted by some fifty snakes. Every python of the *danh-gbi* kind must be treated with respect, and death is the penalty for killing one, even by accident. *Danh-gbi* has numerous wives, who until 1857 took part in a public procession from which the profane crowd was excluded; a python was carried round the town in a hammock, perhaps as a ceremony for the expulsion of evils. The rainbow-god of the Ashanti was also conceived to have the form of a snake. His messenger was said to be a small variety of boa. but only certain individuals, not the whole species, were sacred. In many parts of Africa the serpent is looked upon as the incarnation of deceased relatives. Among the Amazulu, as among the Betsileo of Madagascar, certain species are assigned as the abode of certain classes. The Maasai, on the other hand, regard each species as the habitat of a particular family of the tribe.

Quetzalcoatl depicted as a snake devouring a man, from the Codex Telleriano-Remensis.

The Ancient Egyptians worshiped a number of snake gods, including Apophis and Set, and the Sumerians before them had a serpent god Ningizzida.

In America some of the Native American tribes give reverence to the rattlesnake as grandfather and king of snakes who is able to give fair winds or cause tempest. Among the Hopi of Arizona the serpent figures largely in one of the dances. The rattlesnake was worshipped in the Natchez temple of the sun and the Aztec deity Quetzalcoatl was a feathered serpent-god. In many MesoAmerican cultures, the serpent was regarded as a portal between two worlds. The tribes of Peru are said to have adored great snakes in the pre-Inca days and in Chile the Mapuche made a serpent figure in their deluge beliefs.

Serpent worship was well known in ancient Europe. There does not appear to be much ground for supposing that Aesculapius was a serpent-god in spite of his connection with serpents. On the other hand, we learn from Herodotus of the great serpent which defended the citadel of Athens. The Roman genius loci took the form of a serpent where a snake was kept and fed with milk in the temple of Potrimpos, an old Slavonic god. The ancient Greeks also featured the Gorgons and Medusa in their mythology.

## Fish

According to the Jewish scholar Rashi, the Canaanite god Dagon was a fish god. This tradition may have originated here, with a misinterpretation, but recently uncovered reliefs suggest a fish-god with human head and hands was worshipped by people who wore fish-skins.

Supposedly, there were sacred fish in the temples of Apollo and Aphrodite in Greece, which may point to a fish cult. The goddess at Ashkelon, Atargatis was depicted as half woman, half fish, and according to Xenophon the fish of the Chalus were regarded as gods.

A modern interpretation of Dagon as a "fish-god"

In Japan, there was a deity called Ebisu-gami who, according to Sakurada Katsunori, was widely revered by fishing communities and industries (Qtd. in Naumann, 1974, p. 1). Ebisu, in later traditions, normally appeared in the form of a fisherman holding a fishing pole and carrying a red tai (a perch), but would sometimes take the form of a whale, shark, human corpse, or rock (Naumann, 1974, p. 1). The general image of Ebisu, however, appears to be the whale or the shark, according to Sakurada (Qtd. in Naumann, 1974, p. 2).

During Ebisu-gami festivals, there have been legends told of strange fish creatures which have arrived and been considered sacred. Examples of such fish creatures include familiar species of fish with multiple tails (Naumann,

1974, p. 2). Sometimes these fish were considered to be simply an offering to the deity. Other times, however, they were considered to be Ebisu himself, visiting on the festival day (Naumann, 1974, p. 2).

The praying mantis and the caterpillar, Ngo, are the incarnations of Cagn, a prominent figure in Bushman mythology. It was called the "Hottentots' god" by early European settlers.

The cult of the lizard is most prominent in the Pacific, where it appears as an incarnation of Tangaloa. In Easter Island a form of the house-god is the lizard. It is also a tutelary deity in Madagascar.

## Oracular animals

Animals are frequently used for the purposes of divination. Birds are especially common in this role, as by their faculty of flight they offer themselves to the interpretation as messengers between the celestial and human spheres. Augury was a highly developed practice of telling the future from the flight of birds in Classical Antiquity. The dove appears as an oracular animal in the story of Noah, and also in Thisbe in Boeotia there was a dove-oracle of Zeus. Animal imagery was also often employed in the oracular utterances in Ancient Greece.[5]

In Chinese traditional religion, the tortoise is an oracular animal.

Notable oracular animals of the modern period include Lady Wonder, Punxsutawney Phil, Maggie the Monkey, Lazdeika the Crab, Paul the Octopus and Sonny Wool.

## Shamanism and animals

Animals were an important aspect of the Shaman religion in Central Asia. Also known as “assistant spirits,” “guardian spirits,” and “helping spirits,” animal spirits are an integral part of a shaman's work. The more animal spirits a shaman had under his control, the more powerful the shaman (Waida, 1983, p. 228-229). When a shaman set out to journey spiritually to the outer world, animals were a key component, assisting him in his work. There were three primary reasons for a shaman to take such a journey: to find a lost soul, to bring an animal spirit to the high gods, or to lead a soul to its new resting place in the underworld. All of these were extremely important to followers of shamanism and animals were extremely important in facilitating the shaman's efforts (Waida, 1983, p. 231).

An example of animal spirits in Shamanism comes from the Yenisei Ostiaks culture. During a healing procedure, a shaman invokes a number of animal spirits to help him. The spirits arrive and enter his body. The shaman is not possessed by these spirits; he is free to expel them at any time (Waida, 1983, p. 223). His body begins to leap all over the place, symbolizing that his soul is rising, leaving the earth and going up to the sky. It is a bird spirit that is lifting him through the atmosphere and he cries for it to take him higher so he can see further. According to Adolf Friedrich, at this point the shaman's essence has, in fact, transformed into the bird spirit that crossed the threshold into his body (Waida, 1983, p. 223). He finally spots what he is looking for, the soul of his ill patient. Still assisting him, the animal spirits carry the shaman to the patient's soul. The shaman retrieves it and returns the soul to its rightful place, healing the patient. Without the presence of animal spirits, the shaman could not have accomplished such a feat (Waida, 1983, p. 231).

In the Inner Eurasian religion, the transformation of a shaman's essence into an animal spirit is referred to as “becoming animal” (Baldick 2000, p. 167). The importance of animals in this shamanic religion is shown by the capabilities that animals grant to human beings. Without the assistance of animals, humans from Inner Eurasia were not capable of reaching the sky, traveling rapidly throughout the earth, or going beneath the earth's outer crust, all of which were important activities to the culture (Baldick 2000, p. 167). Heaven was not attainable for a person without the assistance of an eagle. Because of the eagle, an animal, the Inner Eurasians believed that they were capable of achieving their after-life and living in the home of their ancestors and Supreme God after their departure from the earth (Baldick 2000, p. 167). Heaven was represented by the people in assemblies of animals, usually grouped in sevens or nines (Baldick 2000, p. 167). When participating in hunting or warfare, Inner Eurasians also took on animal qualities because they believed it would increase their success (Baldick 2000, p. 167). Animals were a central part of this religion (Baldick 2000, p. 167).

## Religion and animals

### Buddhism

One of the most important sanctions of the Buddhist faith is the concept of ahimsa, or refraining from the destruction of life (Regenstein 1991, p. 234). According to Buddhist belief, humans do not deserve preferential treatment over other living beings. Thus, the world is not specifically meant for human use and should be shared equally amongst all creatures (Epstein 1990). Buddhists recognize that all animals are sentient and are capable of feeling pain, grief, fear, happiness, and hunger (Regenstein 1991, p. 234-235). The Dalai Lama once said "Even ants and other insects will run away from danger... They have intelligence and want to live too. Why should we harm them?" (Qtd. in Regenstein 1991, p. 235). Not believing in inflicting harm on any living, sentient being, some Buddhists also follow a vegetarian diet to avoid causing pain to animals (Regenstein 1991, p. 238).

Avoiding the destruction of life can affect aspects beyond a Buddhist's diet, such as travel plans. In order to avoid crushing any living thing, be it plant, insect, or animal, some Buddhist monks do not travel during rainy seasons (Regenstein 1991, p. 236). Originally, shortly after Buddhism was first founded, monks traveled during all seasons, but public opinion changed this. The people protested that so much life was crushed and destroyed when monks traveled during the wet season. As a result, monks were required to seek shelter during this season and abstain from journeys (Chapple 1993, p. 22).

Living creatures, including humans, culminate to form one large, united life-force in the Buddhist religion. Buddhists, therefore, believe that to harm another living creature is to, in fact, harm yourself as all life-forms are interrelated (Regenstein 1991, p. 237). There are many parables that depict humans sacrificing their lives so that an animal may live. A jataka, or previous incarnation story, tells how the Buddha, (upon hearing the distraught cries of a lioness struggling to feed her hungry cubs), leapt from a cliff and smashed his body to death as an offering, so that she could feed his flesh to them (Chapple 1993, p. 22).

### Jainism

Mahavira, the founder of Jainism, believed that the only way to be released from the cycle of life (birth, death, and then rebirth), one must follow, like Buddhists, ahimsa and not harm any living creature (Regenstein 1991, p. 229). Some Jains will carry a broom with them and sweep their path as they walk to avoid stepping on any living creature. Jains will also wear masks over their mouths to prevent swallowing insects and inspect their fruit for worms. The fruit inspection is not, however, because of their aversion of worms, but for the protection of the worms themselves (Regenstein 1991, p. 229-230). Jains are also only allowed to eat during daylight hours, when their vision is not restricted, so that they avoid eating insects or other small creatures that could possibly be in their food (Regenstein 1991, p. 230).

Jainism includes a lay form which is somewhat less restrictive (Regenstein 1991, p. 231). Basically lay Jains must distinguish between what forms of violence are necessary and unnecessary, but do not have to abstain entirely (Vallely 2002, p. 5). This results in avoiding all forms of hunting, tilling the soil (tilling involved disturbing creatures embedded in the earth), and brewing (brewing involved using living organisms such as yeasts) (Regenstein 1991, p. 231).

Food will never be prepared especially for them. They beg for food from others believing that because the food was prepared for someone else, they are not the cause of violence towards living creatures (Vallely 2002, p. 5).

Lay Jains, who have the financial capacity, will visit animal markets and buy/rescue animals destined for slaughter for the good that it does. (Regenstein 1991, p. 232).

### Hinduism

Hinduism is the primary religion of India (Regenstein 1991, p. 221). Hinduism has evolved over several centuries with vedic times when there was no restriction on animal worship and also animal consumption for food, to later Buddhist and Jain influence that led to the concept of non violence and animal respect and consideration, ahimsa (non-violence) is a major concept in Hindu belief (Regenstein 1991, p. 223). Humans and animals are one family and therefore, humans should treat all living creatures with respect and kindness. Pets are often treated as if they are truly members of the family (Regenstein 1991, p. 223-224).

There are some exceptions to ahimsa in Hinduism. While Hindu belief forbids the slaughter of animals for human sustenance, animal sacrifice was an accepted ritual in *some* parts of India (Regenstein 1991, p. 225). An explanation for this supposed paradox is that a sacrificial animal is not really considered to be an animal, but a symbol. Thus, when the animal is sacrificed, they are sacrificing the symbol and not the animal (Regenstein 1991, p. 226). The killing of an animal for human pleasure or lavishness is prohibited. An example of such lavishness would be hunting for pleasure, a fur coat made from animal skin, etc. (Regenstein 1991, p. 226).

## See also

- Animal sacrifice
- Totemism
- Animism
- Totem
- Moral status of animals in the ancient world
- Animal welfare
- Animal-assisted therapy
- Animal assisted interventions
- Nature worship

## References

[1] Quirke and Spencer 1992, pp. 78, 92–94

[2] Owen, James (2004), "Egyptian Animals Were Mummified Same Way as Humans" (http://news.nationalgeographic.com/news/2004/09/0915_040915_petmummies.html), *National Geographic News*, , retrieved 2010-08-06

[3] Cooper, JC (1992). *Symbolic and Mythological Animals*. London: Aquarian Press. pp. 226–27. ISBN 1-85538-118-4.

[4] "Tiger Culture | Save China's Tigers" (http://english.savechinastigers.org/node/316). English.savechinastigers.org. . Retrieved 2009-03-07.

[5] J. L. Lightfoot, *The Sibylline oracles* Oxford University Press, 2008, ISBN 9780199215461, p. 237, fn. 105 (http://books.google.com/books?id=q5xM-0gWFmYC&pg=PA237&dq="oracular+animal"&hl=en&ei=cuQ1TLKFJ4fEOO6YxOME&sa=X&oi=book_result&ct=result&resnum=1&ved=0CCsQ6AEwAA#v=onepage&q="oracular animal"&f=false).

- This article incorporates text from a publication now in the public domain: Thomas, Northcote Whitbridge (1911). "Animal worship". In Chisholm, Hugh. *Encyclopædia Britannica* (11th ed.). Cambridge University Press.
- Baldick, Julian (2000). "Animal and Shaman: Ancient Religions of Central Asia" New York University Press, New York
- Balfour, Henry (1917). "Some Ethnological Suggestions in Regard to Easter Island, or Rapanui" Folklore, **28**(4).
- Bhattacharyya, Asutosh (1965). "The Serpent as a Folk-Deity in Bengal" Asian Folklore Studies, **24**(1).
- Brown, Theo (1950). "Tertullian and Horse-Cults in Britain" Folklore, **61**(1).
- Chapple, Christopher (1993). "Nonviolence to Animals, Earth, and Self in Asian Traditions" State University of New York Press, Albany
- Epstein, Ronald (1990). "A Buddhist Perspective on Animal Rights" San Francisco State University, (http://online.sfsu.edu/~rone/Buddhism/BuddhismAnimalsVegetarian/Buddhism and Animal Rights.htm)
- Harrell, Stevan; Yongxiang, Li (2003). "The History of the History of the Yi, Part II" Modern China, **29**(3).

- Kindaichi, Kyōsuke (1949). "The Concepts behind the Ainu Bear Festival (Kumamatsuri)", Southwestern Journal of Anthropology, **5**(4), Trans. Minori Yoshida.
- Lantis, Margaret (1938). "The Alaska Whale Cult and Its Affinities" American Anthropologist, New Series, 40(3).
- Livingstone, A (1988). "The Isin "Dog House" Revisited", Journal of Cuneiform Studies, **40**(1)
- Lubbock, John (2005). "The Origin of Civilisation and the Primitive Condition of Man", Kessinger Publishing Company.
- Margul, Tadeusz (1968). "Present-Day Worship of the Cow in India" Numen, **15**(1),
- Meyerowitz, Eva L. R. (1940). "Snake-Vessels of the Gold Coast" Man, **40**.
- Morris, Brian (2000). "Animals and Ancestors: An Ethnography", Berg, New York.
- Naik, T.B. (1958). "Religion of the Anāvils of Surat", The Journal of American Folklore, **71**(281).
- Naumann, Nelly


# Astrolatry

**Astrolatry** is the **worship of stars** and other heavenly bodies as deities, or the association of deities with heavenly bodies. The most common instances of this are sun gods and moon gods in polytheistic systems worldwide. Also notable is the association of the planets with deities in Babylonian, and hence in Greco-Roman religion, viz. Mercury, Venus, Mars, Jupiter and Saturn.

The term **astro-theology** is used in the context of 18th to 19th century scholarship aiming at the discovery of the original religion, particularly primitive monotheism. Unlike *astrolatry*, which usually implies polytheism, frowned upon as idolatrous by Christian authors since Eusebius, astrotheology is any "religious system founded upon the observation of the heavens",[1] and in particular, may be monotheistic.

## Astrolatry

Babylonian astronomy from early times associates stars with deities, but the heavens as the residence of an anthropomorphic pantheon, and later of monotheistic God and his retinue of angels, is a later development, gradually replacing the notion of the pantheon residing or convening on the summit of high mountains.

Sayce (1913) argues a parallelism of the "stellar theology" of Babylon and Egypt, both countries absorbing popular star-worship into the official pantheon of their respective state religions by identification of gods with stars or planets.[2]

Astrolatry does not appear to have been common in the Levant prior to the Iron Age, and becomes popular under Assyrian influence. The Sabaeans were notorious for their astrolatry, for which reason the practice is also known as "Sabaism" or "Sabaeanism". Similarly, the Chaldeans came to be seen as the prototypical astrologers and star-worshippers by the Greeks.

Astrology in the Hellenistic period grew out of Near Eastern and Egyptian practices of astrolatry. Mithraism was a Roman era mystery religion which incorporated many aspects of arcane astral lore derived from Hellenistic astrology.

## Prohibition in Abrahamic religions

The Hebrew Bible contains repeated reference to astrolatry. Thus, Deuteronomy 4:19, 17:3 contains a stern warning against worshipping the sun, moon, stars or any of the heavenly host. Relapse into worshipping the host of heaven, i.e. the stars, is said to have been the cause of the fall of the kingdom of Judah in II Kings 17:16. King Josiah in 621 BC is recorded as having abolished all kinds of idolatry in Judah, but astrolatry was continued in private (Zeph. 1:5; Jer. 8:2, 19:13). Ezekiel (8:16) describes sun-worship practiced in the court of the temple of Jerusalem, and Jeremiah (44:17) claims that even after the destruction of the temple, women in particular insisted on continuing their worship of the "queen of heaven".

Augustine of Hippo criticized sun- and star-worship in *De Vera Religione* (37.68) and *De civitate Dei* (5.1-8). Pope Leo the Great also denounced astrolatry, and the cult of *Sol Invictus*, which he contrasted with the Christian nativity.

The Qur'an contains strong prohibitions against astrolatry.

## Astro-theology

Astrotheology is the study of the astronomical origins of religion; how gods, goddesses, and demons are personifications of astronomical phenomena such as lunar elipses, planetary alignments, and apparent interactions of planetary bodies with stars. Christianity, Judaism, Zoroastrianism, Mithraism, and the ancient Egyptian religions are examples of faiths claimed to be derived from observations of the bodies on the celestial sphere. Examples of deities said to be created as astrological allegories include YHWH, Ra, Horus, Osiris, Mithras, Zoroaster, Helios, Apollo, Lugh, Quetzalcoatl and Jesus Christ.

The term *astro-theology* appears in the title of a 1714 work by William Derham, *Astro-theology: or, A demonstration of the being and attributes of God, from a survey of the heavens* based on the author's observations by means of "Mr. Huygens' Glass". Derham thought that the stars were openings in the firmament through which he thought he saw the Empyrean beyond.[3] The 1783 issue of *The New Christian's magazine* had an essay entitled *Astro-theology* which argued the "demonstration of sacred truths" from "a survey of heavenly bodies" in the sense of the watchmaker analogy. Higginson (1855) argues a compatibility of "Jewish Astro-theology" of the Hebrew Bible, which places God and his angelic hosts in the heavens, with a "Scientific Astro-theology" based on observation of the cosmos.

Manly P Hall (1901-1990) who was a mystic and a 33 degree mason, lectured on astrotheology teaching that each of the three abrahamic faiths has a planet that governs that religion. Judaism is Saturn, the symbol of judaism is a hexagram symbol of Saturn , the day of worship is on saturday day of Saturn. Christianity is the Sun , symbol of christianity is the cross symbol of the Sun ,day of worship sunday day of Sun. Islam is Venus , symbol of Islam the five pointed star and the crescent each symbol is a venus symbol, day of worship is on friday day of venus.

The same term is used by Irvin, Maxwell and Rutajit (2006) in reference to "the earliest known forms of religion and nature worship", advocating the entheogen theory of the origin of religion.

## See also

- Astraea (mythology)
- Astraeus
- Babylonian astrology
- Eosphorus
- Heavenly host
- Heavens (disambiguation)
- Hellenistic astrology
- History of astrology
- Idolatry
- Moon worship

- Nature worship (disambiguation)
- Religious cosmology
- Sabaism
- Sky Father
- Sun worship

## References


[1] OED, citing Derham (1714) as the first attestation of the term.

[2] Archibald Henry Sayce, The religion of ancient Egypt, Adamant Media Corporation, 1913, 237f.

[3] Michael J. Crowe, *Modern theories of the universe: from Herschel to Hubble*, Courier Dover Publications, 1994, ISBN 9780486278803, p. 67.

- William Derham, *Astro-theology: or, A demonstration of the being and attributes of God, from a survey of the heavens*, printed by W. and J. Innys, 1721
- Jan Irvin, Jordan Maxwell, Andrew Rutajit, *Astrotheology and Shamanism*, Book Tree, 2006, ISBN 9781585091072.
- Edward Higginson, *Astro-theology; or, The religion of astronomy: four lectures, in reference to the controversy on the "Plurality of worlds," as lately sustained between Sir David Brewster and an essayist*, E.T. Whitfield, 1855.


## External links

- Jewish Encyclopedia, Star-worship (http://www.jewishencyclopedia.com/view.jsp?artid=1042&letter=S)
- Blackwell Reference Online, Star-Worship (Astrolatry, Sabaism) (http://www.blackwellreference.com/public/tocnode?id=g9780631181392_chunk_g978063118139220_ss1-156)
- Interdisciplinary Encyclopedia of Religion and Science (http://www.disf.org/en/Voci/39.asp)

# Sacred mountains

**Sacred mountains** are central to certain religions and are the subjects of many legends. For many, the most symbolic aspect of a mountain is the peak because it is believed that it is closest to heaven or other celestial bodies.[1] Almost all religions have some sacred mountains - that either are holy (like Mount Olympus in Greek mythology) or are related to famous events (like Mount Sinai in Judaism and descendant religions). In some cases, the sacred mountain is purely mythical, like the Hara Berezaiti in Zoroastrianism. Volcanoes, such as Mount Etna in Italy, were also considered sacred. Mount Etna was believed to have been the home of Vulcan, the Roman god of fire.

## Ancient myths and practices

Mount Olympus.

Mount Olympus is the highest mountain peak in Greece. It was once regarded as the "home of the Greek Gods/The Twelve Olympians of the Hellenistic World". It was also considered the site of the War of the Titans (Titanomachy) where Zeus and his siblings defeated the Titans.

Mount Othrys is a mountain in Central Greece, which is believed to be the home of the Titans during the ten-year war with the Gods of Mount Olympus.

Mount Ida, also known as Mountain of the Goddess, refers to two specific mountains: one in the Greek island of Crete and the other in Turkey (formerly known as Asia Minor).

Mount Ida is the highest mountain on the island of Crete is the sacred mountain of the Titaness Rhea, also known as the mother of the Greek Gods. It is also believed to be the cave where Greek God Zeus was born and raised.

The other Mt. Ida is located in Northwestern Turkey alongside the ruins of Troy (in reference to the Hellenistic Period). The mountain was dedicated to Cybele, the Phyrgian (modern day Turkey) version of Earth Mother. Cybele was the goddess of caverns and mountains. Some refer to her as the "Great Mother" or "Mother of the Mountain". The mythic Trojan War is said to have taken place at Mount Ida and that the Gods gathered upon the mountaintop to observe the epic fight. Mount Ida in Turkey is also represented in many of the stories of Greek author Homer such as Iliad and Odyssey.

Mount Athos, located in Greece, is also referred to as the Holy Mountain. It has great historical connections with religion and classical mythology. In religion it is believed that after the Ascension of the Lord, the Virgin Mary landed on the island and came upon a pagan temple. It was there that the pagan practitioners converted from paganism to Christianity. The Virgin Mary then blessed the land and claimed it her own.[2]

In Classic mythology, Mount Athos is named after the Thracian giant who battled Poseidon, God of the Sea, during the clash of the titans and Gods. It is also said that Greek historian was given the task of creating a canal through the mountain after the failed journey of Persian leader, Xerxes. Overtime, Alexander the Great has become associated with the mountain for his worldly powers. The myth states that Roman architect Dinocrates had wanted to carve Alexander the Great's figure onto the top of the mountain in tribute to him.[3]

The ancient Inca displayed a connection with death and their mountains. It is well known by scholars that the Inca sensed a deep reservoir of spirituality along the mountain range. Situating their villages in the mountains, they felt these places acted as portal to the gods. Ritual child sacrifices called Capachochas were conducted annually, where the most precious gift that could be given (innocent, blemishless, perfect human life) would be sacrificed to the gods. Tremendous effort would be taken as the sacrificial victims would be paraded alive throughout the cities, with multiple festivals and feasts taking place. The final destination would be the tops of some of the highest mountains near their villages, leaving these sacrifices to freeze in the snow. These would take place during great times of distress, during times of famine, violent periods of war, and even during times of political shift. This connection with the mountain as a sacred space is paramount. There would be no other place that would be sufficient or acceptable enough for the gods to accept these gifts. It is neither a surprise nor a coincidence that their honored dead were placed on the highest peaks of the mountains to express the shared connection between the sacred mountain, the gods, and the dead.[4]

## Modern indigenous beliefs

Machapuchare, a sacred Nepalese mountain, viewed from foothills.

Various cultures around the world maintain the importance of mountain worship and sacredness. One example is the Taranaki peoples of New Zealand. The Taranaki center their whole life around the sacred mountain, the Taranaki Mountain. It is no wonder that they shared the same name, as they shared their livelihood from its streams. The rivers that flowed down its steep terrain fed the plants, animals, and gave the tribe all they needed for life. The Taranaki tribe places this mountain into a context of a love story, spelling out the history of their creation in a battle over love, defeat, and a happy ending where this Taranaki Mountain found love with a neighboring volcano. This narrative plays out in the lives of the tribesmen where the mountain is their love, their life force. Life is given from megalith, and when life is taken away, the people are ultimately returned to the mountain. This mountain can be explained as anthropomorphisised, a living organism of its own.[5]

In Korea, people have maintained ancient ways of worshiping mountain spirits. While they are not in fact worshiping the land itself, the gods associated with this worship are united to the land. These spirits are female entities to whom people pay tribute while passing by the mountains, asking for good luck and protection. People also travel to these mountains to ask for fertility. While people generally hold to these female deities for protection or to perpetuate life, one of their most important functions is to protect the dead. The ponhyangsansin is a guardian spirit that is protecting an important clan grave site in the village. Each mountain goddess has an equally interesting story that is tied to their accounts of war against Japan, and the historical legacy of their emperors. Each spirit learned difficult lessons and experienced some sort of hardship. These legacies in the mountains serve as a kind of monument to the history of Korea. While many of the accounts may be true, their details and accuracy are shrouded by time and ritual. While the inaugurations of new ponhyang san sin are not being conducted, fallen important clansmen and leaders are strategically placed in the mountains in order for these strong, heroine-like spirits may fiercely guard their graves. The history of Korea is in turn protecting its own future.[6]

In Japan, Mount Koya-san is the home to one of the holiest Buddhist monastery complexes in the country. It was founded by a saint, Kukai, who is also known as Kobo Dashi and is regarded as a famous wandering mystic; his teachings are infamous throughout Japan and he is credited with being an important figure in shaping early Japanese culture. Buddhists believe that Kobo Dashi is not dead, but will instead awake and assist in bringing enlightenment to all people, alongside the Buddha and other bodhisattvas. It is believed that he was shown the sacred place to build the monastery by a forest god; this site is now the location of a large cemetery that is flanked by 120 esoteric Buddhist temples. Approximately a million pilgrims visit Mount Koya-san a year; these pilgrims have included both royals and commoners who wish to pay their respects to Kobo Dashi. Mount Fuji is another sacred mountain in Japan. Several Shinto temples flank its base, all to pay homage to the mountain. The common belief is that Fujiyama is the incarnation of the earth spirit itself. The Fuki-ko sect maintains that the mountain is a holy being, and the home to the goddess Sengen-sama. Annual fire festivals are held here in her honor. Fujiyama is also the site of pilgrimages; reportedly 40,000 people climb up to its summit every year.[7]

Primarily located in Bhutan, the Himalayas have classically been referred to as the home of gods and immortals. Buddhist and Hindu tradition both hold the Himalayas to be the center of the world, and they maintain that the Himalayas hold the underworld, earth, and heavens together.

In some cultures, Mount Kailash is referred to as the "center of the universe" and is considered to be the resting place of Gods. Tibet's Mount Kailash is a sacred place to four religions: Buddhism, Jainism, Hinduism, Bon Po (a native Tibetan religion prior to Buddhism), and Ayyavazhi religions. According to Hindu tradition, Kailash is the home of the deity Shiva. Some sects hold that Kailash is paradise and the final destination of all souls in the world.In Hindu religion, Mount Kailash also plays an important role in Rama's journey in the ancient Sanskrit epic, Ramayana. Buddhists hold that Kailash is the home of Samvara, a guardian deity, and a representation of Buddha. Buddhists believe that Mount Kailash has supernatural powers that are able to clean the sins of a lifetime of any person. Followers of Jainism believe that Kailash is the site where the founder of Jainism reached enlightenment. Bon Po teaches that Kailash is the home of a wind goddess.

Mount Meru is a cosmic mountain which is described to be one of the highest points on Earth and is the center of all creation. In the Hindu religion, it is believed that Meru is home to the god Brahma, who is believed to be the father of the human race and all the demigods produced afterward. Indian cosmology believes that the sun, moon, and stars all revolve around Mount Meru. Folklore suggests the mountain rose up from the ground piercing the heavens giving it the moniker "navel of the universe".[8]

According to the Torah, and consequently the Old Testament of the Bible, Mount Sinai is the location that Moses received the Ten Commandments directly from God. The tablets form the covenant, which is a central cornerstone of Jewish faith. Saint Catherine's Monastery is located at the foot of Sinai. It was founded by empress Helena, who was the mother of the first Christian Roman emperor, Constantine. It was completed under the rule of Justinian two centuries later. The monastery was visited by the prophet Muhammed, who blessed it and promised "that it would be cherished by Muslims for all time".[9] Today, the monastery is home to a group of Greek Orthodox monks, as well as a large collection of Byzantine art, illuminated manuscripts, icons, and books; the collection of icons in particular has been proclaimed one of the oldest in the world.

The Navajo possess a strong belief system in regards to the natural-supernatural world and have a belief that objects have a supernatural quality. For example, the Navajo consider mountains to be sacred. There are four peaks, which are believed to have supernatural aspects. The mountains each represent a borderline of the original Navajo tribal land. The mountain ranges include Mount Taylor, the San Francisco Peaks, Blanca Peak, and Hesperus Peak located in the La Plata Mountains.

Each mountain/peak is representative of a color, direction, and correlates with a cultural light phenomenon dealing with the cosmic scheme of the rising and of the setting sun. Directionally, the mountains are described in a clockwise motion following the movement of the sun beginning with the eastern mountain of Blanca Peak. Blanca Peak is associated with the color white and the "Dawn Man" referring to the rising of the sun. Next in the south is Mt. Taylor, which is associated with the color blue and the "Horizontal Blue Man" referring to the daytime. In the west is the San Francisco Peaks, which is representative of the color yellow and the "Horizontal Yellow Woman" and is associated with the setting of the sun. And finally in the north is the Hesperus Peak of the La Plata Mountains which is given the color black and belongs to the light phenomenon of the "Darkness Woman" representing the nighttime.[10]

## Community identity

History shows that mountains were commonly part of a complex system of mountain and ancestor worship. Having immortalized fallen brethren in the edifice, the people share a common allegiance with all the other people of a community. The meanings that were etched into the mountain and mound terrain connected the villagers. They were all subject to the same landscape and village history, which were bound together by their cultural significance. The history of ancestors could be told by simply pointing at specific mountains and remembering the stories that were passed down throughout the generations. The worship of ancestors and the mountains were largely inseparable. An interconnected web between history, landscape, and culture was thus formed.[11] Examples of this would be the Hindu belief that Mount Kailas is the final resting place for the souls of the dead, as well as the large cemetery placed on Mount Koya-san.

Sacred mountains can also provide an important piece of a culture's identity. For example, Messerli and Ives write, "The Armenian people regard Mount Ararat, a volcano in eastern Turkey believed to be the site of Noah's Ark in the Bible, to be a symbol of their natural and cultural identity".[12] As a result of the mountain's role as a part of a cultural identity, even people who do not live close to the mountain feel that events occurring to the mountain are relevant to their own personal lives. This results in communities banning certain activities near the mountain, especially if those activities are seen as potentially destructive to the sacred mountain itself.

## Pilgrimages

To date, Kailas has never been climbed, largely due to the fact that the idea of climbing the mountain is seen as a major sacrilege. Instead, the worshipful embark on a pilgrimage known as the kora. The kora consists of a 32 mile path that circles the mountain, which typically takes five days with little food and water. Various icons, prayer flags, and other symbols of the four religions that believe Kailas is sacred mark the way. To Buddhists and Hindus, the pilgrimage is considered a major moment in a person's spiritual life. Olsen writes, "One circuit is believed to erase a lifetime of sin, while 108 circuits is believed to ensure enlightenment".[13]

As one of the most sacred mountains in the Middle East, and one of the few sacred places arguably equally revered by Jews, Christians, and Muslims, Mount Sinai is a popular location for pilgrimages. One route, called the Steps of Repentance, is composed of 3,000 steps that were carved into the mountainside by a monk. Various sights mentioned in the Old Testament can be seen on the mountain's summit, such as the area where Moses "sheltered from the total glory of god".

Sacred Mountains are often seen as a site of revelation and inspiration. Mount Sinai is an example, as this is the site where the covenant is revealed to Moses. Mount Tabor is where it is supposed Jesus was revealed to be the son of god. Muhammed is said to have received his first revelation on Mount Hira.[14] The mountains' roles as places of revelation and transformation often serve to attract tourists as much as they do religious pilgrims. However, in some cases, the financial revenue is overlooked and sacred mountains are conserved first due to their role in the community.[15]

## Conservation

Views of preservation and sacredness become problematic when dealing with diverse populations. When one observes the sacred mountain of the Sacramento Valley in the United States, it becomes clear that methods and opinions stretch over a vastly differing body of protesters. Shasta Mountain was first revered by the Native American tribe, the Wintu. Shasta was in effect a standing monument for the individuals of their cultural history. This bounded view of sacred mountains changed drastically during the 1800s. It is commonly assumed that sacred mountains are limited by a single society, trapped in a time capsule with only one definition to explain it: the indigenous tribe. Shasta's glory had expanded to multiple regions of the world, communities of differing religions making their pilgrimage up to the summits of this glorious mountain. The Wintu tribe did not hold a monopoly on the sacredness anymore. There were others contesting to the meanings, adding new rituals and modifying old ones. With the advent of new technology and desires to turn this mountain into a skiing lodge, angry voices from all over the world rose up with variants of demands on why and how we should preserve this beautiful mountain.

Almost every day different religious practices such as nude bathing, camping out with magic crystals, yoga, and many "quasi-Christian" groups such as the I AM march their ways up to the tips of this mountain. With this activity the mountain pathways become clustered, cluttered and littered. Even the pathways' existence leads to erosion, and further slow degradation of the mountain. The Wintu tribe has voiced concerns and asked for support from the government to regulate the activities practiced on "their" mountain saying that "they are disturbed by the lack of respect" shown for this piece of land. It has become greatly debated if the more vulnerable and "spiritually desirable" places of the mountain should be closed and maintained only by the Wantu tribe, who see this land as a sacred graveyard of their ancestors, or open to all who seek spiritual fulfillment such as the modern day group of the I AM.[16]

Sacred mountains are often viewed as the source of a power which is to be awed and revered. Typically, this power is seen as dangerous, with the requirement that it be treated carefully and with respect. Often, this means that access to the sacred mountain is restricted. This could result in climbing being banned from a sacred mountain completely (as in the case of Mount Kailas) or for secular society to give the mountain a wide berth. Because of the respect accorded to a mountain's sacred power, many areas have been declared off limit for construction and remain conserved. For example, a large amount of forest has been preserved due to its proximity to Mount Koya-san. Additionally, sacred mountains can be seen as the source of something vital. This could be a blessing, water, life, or healing. Mount Kailas's role as the source for four major rivers is celebrated in India and not simply seen as mundane. Rather, this also adds to its position as a sacred place, especially considering the sacred position of the Ganges river in Indian culture. Mountains that are considered home to deities are also central to prayers for the blessings from the gods reputed to live there. This also creates a sense of purity in the source of the mountain. This prompts people to protect streams from pollution that are from sacred mountains, for example.

## Sacred mountains

- Áhkká - regarded by the Sami people as a holy mountain.
- Mount Aqraa (Zaphon)
- Mount Ararat - alleged by some to be the site of Noah's ark and holy to the Armenian Apostolic Church.
- Mount Athos - also known as the *Holy Mountain*, Greece
- Black Hills
- Burkhan Khaldun - Khentii Province, Mongolia
- Mount Everest
- Mount Fuji
- Mount Gerizim - as claimed taught to be the location of the Holies of Holies by God to the Samaritans.
- Mount Kailash, sacred to Hinduism, Buddhism, Jainism and Bön.
- Laoshan
- Mauna Loa/Mauna Kea- volcanic eruptions were thought to be a result from the Hawaiian Goddess of fire Pele (deity) when in argument with her siblings[17]
- Machu Picchu, Huayna Picchu, and other mountains were sacred to the Inca locals.
- Nanda Devi - also known as Bliss-Giving Goddess, Republic of India/Greater Nepal
- Mount Makiling, Mount Arayat, and Mount Lantoy, of the Philippines, and their protectors, Maria Makiling being the protector of Mount Makiling
- Mount Banahaw, Mount San Cristobal- The Holiest place in the Philippince, Termed as the yin and yang mountain.
- Mount of Olives
- Phnom Kulen
- Mount Shasta
- Sulayman Mountain
- Teide - instead of creating the world for aboriginal guanches, Spain
- Uluru - also known as Ayers Rock, Australia
- Mount Vesuvius
- Mount Zion together with **Mount Sinai** and **Mount Ararat** the sacred mountains of the Tewahedo tradition.

## See also

- Sacred Mountains of China
- World mountain

## External links

- Sacred Mountains [18]
- Heilige Berge [19] (German)
- Sacred Mountains and *Sanshin* mountain-spirits of Korea [20]

## References


[1] Naess, Arne.Mountains and Mythology. Trumpeter 1995

[2] Mount Athos: Renewal in Paradise By Graham Speake. New Haven and London: Yale University Press, 2002. 294 pages. ISBN 0-300-09353-5

[3] Dora, Veronica Della. 2005.Alexander the Great's Mountain. Geographical Review Vol. 95 Issue 4 pg. 489.

[4] Ceruti, Constanza.2004 Human Bodies as Objects of Dedication at Inca Mountain Shrines (North-Western Argentina). World Archaeology 36:103-122.

[5] Smith, Ailsa. 1997. Who is that Mountain Standing there... Its Taranaki... In Mountains of the World: A GLobal Priority. B. Messerli and JD Ives, eds. Pg 57. Informa Health Care.

[6] Grayson, James H.1996 Female Mountain Spirits in Korea: A Neglected Tradition. Asian Folklore Studies 55: 19-134.

[7] Olsen, Brad. Sacred Places Around the World: 108 Destinations. 2nd ed. San Francisco: CCC Publishing, 2008. 34-83. Print.

[8] Mabbett, I.W. 1983. The Symolism of Mount Meru. History of Religions Vol. 23 No.1 Pg. 64-83

[9] Olsen, Brad. Sacred Places Around the World: 108 Destinations. 2nd ed. San Francisco: CCC Publishing, 2008. 34-83. Print.

[10] Lamphere, Louise. 1969.Symbolic Elements in Navajo Ritual. Southwestern Journal of Anthropology Vol. 25 No. 3 pg. 379-305

[11] Dillehay, Tom D. 1995 Mounds of social death: Araucanian funeraruries and political succession. In Tombs for the Living: Andean Mortuary Practices. T. Dillehay Ed. pp. 281-314 Dumbarton Oaks Washington, DC

[12] Messerli, Bruno, and Jack D. Ives. Mountains of the World: A Global Priority. 17. International Mountain Society, 1997. 39-54. Print.

[13] Olsen, Brad. Sacred Places Around the World: 108 Destinations. 2nd ed. San Francisco: CCC Publishing, 2008. 34-83. Print.

[14] Quran Ch. 96

[15] Messerli, Bruno, and Jack D. Ives. Mountains of the World: A Global Priority. 17. International Mountain Society, 1997. 39-54. Print.

[16] Huntsinger, Lynn and Maria Fernandez -Martinez 2000 Spiritual Pilgrims at Mount Shasta California. Geographical Review 90: 536-558

[17] Hawaiian Legends of Volcanoes By W.D. Westervelt ISBN 978-0559082283

[18] http://www.sacred-destinations.com/sacred-sites/sacred-mountains.htm

[19] http://www.emmet.de/hb.htm

[20] http://www.san-shin.org/

# Article Sources and Contributors

**Nature worship** *Source*: http://en.wikipedia.org/w/index.php?title=Nature_worship *Contributors*: BD2412, Beno1000, Cwmhiraeth, Dbachmann, Kee211, Luckas Blade, NTK, Neelmack, Opus88888, Silvonen, USchick, Wiki-uk, , 11 anonymous edits

**Nature** *Source*: http://en.wikipedia.org/w/index.php?title=Nature *Contributors*: $yD!, 119, 130.94.122.xxx, 161.53.43.xxx, 203.109.250.xxx, 5alam83, Aarchiba, Abdullais4u, Addihockey10, AdelaMae, After Midnight, AgentPeppermint, Ahoerstemeier, Ainlina, Aitias, Ajamiesonnz, Ajnem, Akcwikiaccount, Alansohn, Alison, All Is One, Allens, Alxeedo, Amalas, Amikeco, Anakin101, AnakngAraw, Ancheta Wis, Anclation, Andres, Andrew Lancaster, Andy Marchbanks, Andycjp, Angela, Animum, Anon30000, Antandrus, Anthere, ApersOn, Aphaia, Arakunem, ArbiterOne, Arolga, Arthur Rubin, Ashley Y, Atbirder, AutoGyro, Awickert, AxelBoldt, Aymatth2, AzaToth, BD2412, BGormley100, BSATwinTowers, Bagworm, Banno, Bareja gulshan, Bart133, Basketball110, Bdiscoe, Ben-Zin, Bensaccount, Bharat muchhala, Bidiot, Bijl0130, Bkonrad, Blanchardb, Blue520, Bobo192, Bongwarrior, Borksamoht, Brandon, Brian0918, BrianBrian123, Burlywood, CALR, COMPFUNK2, CRRaysHead90, CalumH93, Canderson7, Cantus, CapitalR, Captain-tucker, Carinemily, Carlo.Ierna, Celloman89, Cenarium, Cessator, Chamal N, Chocolateee, Chun-hian, Clear8, Cmichael, CommonsDelinker, Contre-boutant-ex, Conversion script, Correogsk, Css, Cybercobra, DARTH SIDIOUS 2, DVD R W, DancingPenguin, Daniel Case, Daniel Collins, Danjammil, Dark hyena, Dave souza, David Little, Deglr6328, Demmy, Dendodge, Derek Ross, Didora, Dilaittan, Dina, Djr32, Dpark, Dpv, Dr.Bastedo, Dralwik, Driftwoodzebulin, Druon, Duffman, EagleKnowledge13, Ed g2s, Edward, Eleassar777, Ems2715, EncMstr, Encyclopedia a-z, Epbr123, EricR, Escape Orbit, EugeneZelenko, Everyking, Explicit, Fastily, Faturita, Fayenatic london, Fdhs, Fffaar, Fieldday-sunday, Fifth Fish Finger, Fir0002, Firepink, Flewis, Fly by Night, Foxandpotatoes, Fragilesoul25, Francs2000, FreplySpang, Fuhegikfgyowiuegf, FullMetal Falcon, G11221, GKaczinsky, Gadfium, Gaf.arq, Gaius Cornelius, Gavrilis, Gdarin, Gdo01, Gdr, Gene Nygaard, Gettingtoit, Glacialfox, Glen, Go for it!, Gogo Dodo, Graham87, Guanaco, Gurch, Halio, Hard Raspy Sci, He Who Is, Headbomb, Henning Makholm, Herakles01, Hertz1888, Hiddekel, Hordaland, Hotcolla, Husond, I8bozo, Ian Pitchford, Icairns, Icseaturtles, Ike9898, Indosauros, Infinity0, Infrangible, Iris92, ItsZippy, J. Finkelstein, JJ Harrison, JJJJS, JNW, JamesBWatson, Jasonmrazlover, Jasonzhuocn, Java4, Jcbutler, Jeepday, Jeffq, Jfkmslol, Jlencion, Joanjoc, JoanneB, JoeC 4321, JoeSmack, John, Johnbibby, Jon186, Jose77, Josephprymak, JoshuaZ, Jpandigamage, Jsayre64, Jstlerce, Jtneill, Jujutacular, Jusdafax, Jusjih, K, KF, KLooove, Kagredon, Karol Langner, Kelovy, Kenny sh, Khalid Mahmood, Kier07, Kinkify, Kirrages, Kiwifun, Kku, Kleinhev, Koavf, Krea, Ktsquare, Kusunose, LOL, Langdell, Lauriec, Ldonna, Leighxucl, Levineps, Lifung, Lightmouse, Lionel711, Lir, Logologist, Loldogslol, Lolololol12345678, Lulubell77, Lumos3, Lupin, Lycurgus, MER-C, MONGO, MPF, Malleus Fatuorum, MarSch, MarsRover, Martpol, Materialscientist, Matt j fox, Maurreen, Mav, Mayooranathan, Mbz1, McGeddon, Medeis, Meggar, Mejor Los Indios, Mel Etitis, Mentifisto, Mepolypse, Metamagician3000, Mgiganteus1, Michael Hardy, MichaelMaggs, Midnight-snow, Mild Bill Hiccup, MithrandirAgain, Mlouns, Modernist, Moneybagbrian55, MrBell, Munita Prasad, Mursel, Mxn, NYC Writes, Nakon, Narayanese, Natcase, Naturalist Jim, Nature.religion, NatureA16, Naveen Sankar, Nectarflowed, Neko85, Nepaheshgar, Newbyguesses, Nick Number, Nimblneicky1, Nixeagle, Nk, Noetica, Nordpesty, Norm, Northamerica1000, Nzelnick, Oceanuts, OhanaUnited, Okedem, Oliver202, Omegatron, Onco p53, One-world-generation, Oxymoron83, PPdd, PTJoshua, Paine Ellsworth, Papa November, Patientone, Patricusrex, Paulbalegend, Peace for All, Peak, PericlesofAthens, Peter, Peterbop, Petiatil, Petrvs, Phantomsteve, Phenz, Philg88, Philip Trueman, PhnomPencil, Phoenix-forgotten, Phr0gor, Pinethicket, Pink!Teen, Plasticup, Pokrajac, Poor Yorick, PoorLeno, Qmwne235, Quadell, Quantum4 (usurped), Quarma, Quintote, Qxz, R Lee E, RDF, RJHall, RL0919, RadioFan, Raven4x4x, Reach Out to the Truth, Reaper Eternal, Redheylin, Reinis, Resident Mario, Rgbroitman, Rich Farmbrough, RichardF, Rjwilmsi, Rmky87, Roadnottaken, Roberta F., RockGod166, RodC, Rosi Crane, Rowena09, RoyFocker, Rstafursky, Ruud Koot, RyanParis, Saifani, Sango123, Scheinwerfermann, Scohoust, Scope creep, SeanMack, Seresin, ShakingSpirit, ShelfSkewed, Shenme, SiobhanHansa, Skizzik, Skywalker, Smalljim, Smartmannnm, Smartse, Snixon, Snowytrix, Solipsist, Son nikko, SorbyRock, Soundwatershed, Southern Illinois SKYWARN, Special-T, SpuriousQ, Stalefries, Stbalbach, Stemonitis, Stephenb, Steveaustin24, Steven Weston, Stevertigo, Stickee, Strife911, Sunray, Svick, Synergism, TYelliot, Tabletop, Tanvir Ahmmed, Taxman, Taylors1, Tazmaniacs, Tbhotch, Tempestangelb, Tempodivalse, Tevildo, The Anome, The Evil Spartan, The Hybrid, The Obfuscator, The Thing That Should Not Be, The undertow, Thingg, Thljcl, Tide rolls, TimVickers, Timir2, Tpbradbury, Triwbe, Trulyelsa, Twinxor, Tylerlouden12, Ulric1313, Umofomia, Uncle Dick, VBGFscJUn3, Valley2city, Van helsing, Vanished User 0001, Vanished User 4517, VanishedUser314159, Vanka5, Vig vimarsh, Vince, Voltteri, Voyagerfan5761, Vsmith, Walkerma, Walton One, Waxinggibbous999, Whispering, Wifliboi, WillowW, Winchelsea, Wine Guy, Woohookitty, Wsduvall, XJamRastafire, Xp54321, Xslf, Ybk33, Ynhockey, Yoasif, Yoninah, Yoshi348, Zacharie Grossen, Zeratul1212, Ævar Arnfjörð Bjarmason, 858 anonymous edits

**Cosmos** *Source*: http://en.wikipedia.org/w/index.php?title=Cosmos *Contributors*: A930913, Aethralis, Alansohn, All Is One, Allraoul, Altenmann, Ancientmacedon, Andre Engels, Arcendet, AtmanDave, AxelBoldt, Bamos, Being blunt, Bigwyrm, Bkonrad, Bogey97, BostonMA, BruceDLimber, Brumi, Bryan Derksen, Burntsauce, C1k3, CalicoCatLover, Carnby, Ceyockey, Cgmusselman, Chameleon, ChangChienFu, Chaos, CommonsDelinker, Copymates, Cybercobra, Cynwolfe, Cyrius, DJProFusion, Daniel Mietchen, Darknext, David R. Ingham, Dazedbythebell, Dbachmann, Dialectric, Dna-webmaster, Don Braffitt, Doulos Christos, Dust Filter, DwightKingsbury, EME44, Eb.hoop, El C, Eleuther, Eloquence, Facius, Falcon8765, Faramir1138, Fconaway, Feeeshboy, Feelin420247, Felipe P, Filemon, Fizan, Foobaz, GB fan, Gene s, Gianfranco, Goethean, GrahamN, Grendelkhan, Greyhood, Headbomb, Hiltonbooks, IPSOS, Iam on andromeda, Is Mise, IvanLanin, JJ Harrison, JMK, Jeffloveswaffles, Jimp, Kariteh, Keraunos, Kku, Kostisl, Kungfuadam, Lepota, Liquorthief, Luk, Macedonian, Markmark12, Maurice Carbonaro, MaxErdwien, Mbell, MelbourneStar, Merovingian, Michael Hardy, Monkeykiss, Mr Stephen, Mr.chewy, Muke, Music Sorter, Nagelfar, Nataly8, Newone, NexCarnifex, NickelShoe, Nlu, OP8, Olmer, Oneinfinitemind, Pjacobi, Podzemnik, RDF, RJBurkhart3, RandomCritic, Ras, Raven in Orbit, Reaverdrop, Redheylin, Retiono Virginian, Saberlotus, Sam Korn, SheikYerBooty, Siddharth Prabhu, Sj, Smilo Don, SoCalSuperEagle, Stepa, SteveMcCluskey, Strait, Susyr, TeleComNasSprVen, Template namespace initialisation script, Theathenae, Thumperward, Tide rolls, Tokle, Trevor MacInnis, Tyrenius, Vanished User 0001, VanishedUser314159, Versageek, Viriditas, WHEELER, Weetoddid, Wertuose, Yidisheryid, Zorio, ملاع بوبحم كيب يلعنابرق, 136 anonymous edits

**Universe** *Source*: http://en.wikipedia.org/w/index.php?title=Universe *Contributors*: -Midorihana-, 007david, 03ctodd, 129.128.164.xxx, 271828182, 2over0, 8732Spacefish, A-giau, ABF, AGK, ALovelyOne1, Aardark, Abdullais4u, Abhiag, Abhinandan27, Abob6, Abomasnow, Acjelen, Adambro, Adamlewis157, Addshore, Adi, AdjustShift, Adric Kearney, Aeonx, Aeosynth, Aetheling, Ageekgal, AgentPeppermint, Agge1000, Agüeybaná, Ahoerstemeier, Aitias, Aj767, Ajp4, Aka042, Akhil.aggarwal2, Akshat2, Alan268, Alansohn, Alasdair Routh, Albert.white, Aldaron, Aldebaran66, Aldie, Ale Ronzani, Alecsdaniel, AlexiusHoratius, Alfie66, Ali, Alicam8, AlimanRuna, All Is One, AlphaEta, Altenmann, Amicaveritas, Amirazemi, Amphibio, AnakngAraw, Andonic, Andres, Andromeda321, Androo, AndrooUK, Andycjp, Angela, Animonster, Animum, Anna Lincoln, Anonymous Dissident, Anonymous editor, Antandrus, Anubad95, Apeconmyth, Apparition11, Ardric47, ArielGold, Army1987, Art LaPella, Art and Muscle, Arthur Rubin, Ascidian, Astronautics, Athenean, AuburnPilot, AugPi, Aunt Entropy, Autonova, Awesome streak, Awickert, Awolf002, AxelBoldt, AzaToth, Azazeel, Azcolvin429, B15nes7, B9 hummingbird hovering, Baaleos, Babebait227, Badacmw90, Bangarangmanchester, Barbara Shack, BarretB, Barticus88, Barts1a, Bbatsell, Bblackmoor, Beej175560, Beetstra, Bejnar, Beland, Ben Ben, BenRG, Benjamin Nicholas Johnston, Bensaccount, Berkeley626, Beyond My Ken, Bhawani Gautam, Bibleboy14, Bigfatmonkeyturd, Bilbo571, Bilbobee, Bjoern, Blah987654, Blanchardb, Blightsoot, Bloodredrover, Bluto7, Bob Andolusorn, Bobblewik, Bobianite, Bobo192, Bobsexual, Bonadea, Bongwarrior, Boogawooga02, BoomerAB, Borislav, Boud, Brainmachine, Brave warrior, Breadhead23, Breakfastchief, Brest, Brian0918, BrianGV, Brianga, Brooktree, Bryan Derksen, Bsadowski1, Bt8257, Bubba hotep, Bubble queen, Buchanan-Hermit, Bullzeye, Bulwersator, Butros, Butterfly 25007, CYD, Caco de vidro, Cadiomals, Call me Bubba, Calmer Waters, Caltas, Canadian-Bacon, Canderson7, Canterbury Tail, Canuck100, Canuckle, Capricorn42, Captain-n00dle, Captain-tucker, CardinalDan, Catgut, CatherineMunro, Cenarium, Ceris2, Chanting Fox, Chaos, CharlesC, Cheese12345cheese, Chetvorno, Chochopk, Chowbok, Chretienorthodoxe, Chrisbrl88, Christopher Thomas, Ciotog, Citationeeded, Citizen Premier, Ckatz, Clarince63, Claus Ableiter, Cloroplast horse, Closedmouth, Coldplayrock08, CommonsDelinker, Compdude47, Confession0791, ConfuciusOrnis, Conversion script, Cool Blue, Cordell, Coreylook, Cosmic Latte, Cranberry5553, Craner Murdock, Crazyj922, Cuckooman4, CuteHappyBrute, Cybercobra, Cytocon, DARTH SIDIOUS 2, DHN, DJ Clayworth, DVD R W, DVdm, DW40, Dalekian, Dang3210, Danger, Daniel Case, Danielrcote, Dannyat43, Dantesparda271191, Dantheman2008, DarkElrad, DarkLouis, Darkowlf8592, Darren23, DarrynJ, Darth Panda, Dath Dath Binks, Davewho2, Davewild, DavyJonesGSB, Dazedbythebell, Dbachmann, Dbannie07, Dbenbenn, Dchristle, Dcs002, DeadEyeArrow, Deadly Matty, Deadman3215, Debashish Mahapatra., Debresser, December21st2012Freak, Deconstructhis, Deepvalley, Delldot, Deltabeignet, Denisarona, Denton22, DerHexer, Dfrg.msc, Diannaa, Dionyseus, Dirtyharry2, Discospinster, Diyako, Dna-webmaster, Doczilla, Domi33, Donthegon101, DorJ, DoubleBlue, Dougweller, Dr A Thompson, Dr. Sunglasses, Dr.K., DrOxacropheles, Drbogdan, Dreadstar, DreamOfMirrors, Drmies, Drooling Sheep, Dsewell, Dub8lad1, Dude00311, Dumbdarwin, Dupz, Dv82matt, Dycedarg, Dylan Lake, EEPROM Eagle, EMC125, ESkog, Eb.hoop, Echtner, Ed Poor, Eddie Dealtry, EdgeOfEpsilon, Eekerz, Eequor, Ekrub-ntyh, El C, Elassint, Elert, Enviroboy, Eog1916, Epbr123, Epeefleche, Epipelagic, Erdelyiek, Erebus Morgaine, EricWaltonBall, Erielhonan, Err0neous, Escape Orbit, Eu.stefan, Eusticeconway, Evercat, Everyking, Evil saltine, Evlekis, Excirial, Exodio, Exploreuniverse, FKmailliW, Fabrictramp, Falcon8765, False vacuum, Familyguylover64, Faradayplank, Farheen1973, Faus, Favonian, FayssalF, Fbs. 13, Felixmonk1, FellowWikipedian, Fences and windows, Fetchcomms, Firetrap9254, Firewall62, FlamingSilmaril, Floria L, Fluffernutter, Fratrep, Frazzydee, Freakofnurture, Fredrik, Friedpotatoeparty, Friendly Neighbour, Fritzpoll, Frosty0814snowman, Funandtrvl, Furkhaocean, Furrykef, Futurebird, Futuristcorporation, Fvw, Fyrael, G. Campbell, GSMR, GSlicer, GWPFBE, Gadfium, Gaff, Gafferuk, Gail, Gaithdroby500, Gakon5, Gakrivas, Galor612, Gandalf61, Gap9551, GcSwRhIc, Gdm, Ged UK, Geniustwin, Geogeogeo, Geometry guy, George585, Georgiocj, Ghostillnses, Giftlite, Gifðas, Gilderien, Gilliam, Git2010, Glacialfox, Glenn, Godzilla 2002, Gogo Dodo, Golbez, GomeonaFinnigan, Goodvac, GorillaWarfare, Graeme Bartlett, GraemeL, Grafen, Graham87, Grahamec, Gregbard, GregorB, Grim23, Grubtatorship, Guettarda, Gunfists, Gurch, Gurchzilla, Guy Peters, Gwark, Gwen Gale, Gwernol, H1tchh1ker4, Hahaho, HamburgerRadio, Hans Adler, Haoie, Hardyplants, Harmi.banik, HarryMcDeath, Hazmat2, Hdt83, Headbomb, Hectorthebat, Hellopunish3r, HenryLi, Herbee, Hermanschrader, Hermes the Wise, HexaChord, HiEv, Hibernian, Hihihihibye, Honeycake, Hontogaichiban, Hotturbos, Houdabouda123, Hu12, Hurdygurdyman1234, Hurricanefloyd, Hut 8.5, Hve, Hyad, Hydnjo, HyperSonic X, Hypnosifl, I LIVE 2 TROLL, IGeMiNix, II MusLiM HyBRiD II, ILSS, IVAN3MAN, Iamnotanorange, Iamorlando, Ian Dalziel, Icairns, IceUnshattered, Ihatejustinbeiber, Ilikepie2221, Imstillhereyoh, Infinitesolid, Inkpot80, Insanephantom, Inswoon, InternetMeme, Iridescent, Irishguy, IronGargoyle, Island Monkey, Ixfd64, J Di, J.delanoy, J04n, J8079s, J991, JEREMYBB, JFHJr, JForget, JJ Harrison, JRSpriggs, JWSchmidt, Jacek Kendysz, JackizCool69, Jackol, JacobParker100, Jagged 85, Jaimeescobar, Jake Wartenberg, Jamiepgs, JavierMC, Jawed, Jaxl, Jbvjkhgvkjukjhvkj, Jcobb, Jdorwin, JedediahBaugh, Jeff G., Jeffo223344, Jeffrey Mall, Jeffrey O. Gustafson, Jeglikerdeg, Jeltz, JeremyA, Jesuschristlover, JetLover, Jhacob, Jhon montes24, Jim1138, JimVC3, JimWae, Jimmyjkjr, Jimothytrotter, Jimp, Jinsubpretzels, Jj137, Jkveton08, Jleon, JoanneB, Joe Jarvis, Joel grover, Joelholdsworth, Joernderschlaue, John, John D. Croft, John haley, JohnCD, Johnny4netglimse, Johnstone, Johntex, Jok2000, Joke137, Jolekweatin, Jonkerz, Jordancpeterson, Jorfer, JorisvS, Jorunn, Jose77, Joseaperez, Joshua098, Jovianeye, Jpabian01, Jpk, Jpmccord, Jrossr, Jstana, Jules93, Jumpingjacksparrow, JuniorMuruin, Jusdafax, Just James, JustKiddingPro2, Justintan88, Juzeris, Jzhuo, Jéské Couriano, K, K. Annoyomous, K0hlrabi, KGasso, Ka Faraq Gatri, Kaldari, Kamakazii101, Karl-Henner, Karol Langner, Kasyap.d, Katalaveno, Katherine, Kazvorpal, Kbev, Keilana, Kel-nage, Kentgen1, Keraunos, Kevin B12, KibaKibbles, Kicking222, Kiefer100, Killiondude, King Pickle, King of Hearts, Kingpin13, Kip the Dip, Kipala, Kirstylovesatl, Kjgiant, Knightshield, KnowYourCosmos, Knowledge Seeker, KnowledgeOfSelf, Kntrabssi, Koavf, Kookyunii, Kotjze, Kozuch, Kreachure, Krellis, Krich, Kubigula,

Kukini, Kurochka, Kuru, L Kensington, LAS1180, LORDGOD7777, La Pianista, Lambiam, Landon1980, Larry Lawrence, Latitude0116, Lawnchair On Jupiter, Lcarscad, LeaveSleaves, Lee Daniel Crocker, Leemadd92, Legare, LeighvsOptimvsMaximvs, Lemmey, Leszek Jańczuk, Libertype, Liftarn, Light-assasin, Lightdarkness, Lightmouse, Lilbrew369, Lindberg, Lir, Llort, Logan, Logan6362, Lokal Profil, LonelyMarble, Looie496, Looxix, Loren.wilton, Lradrama, Lrrasd, LucChickenfingers1873, M karthikkannan, M4gnum0n, MER-C, MIT Trekkie, MONGO, MPerel, Mac, Machine Elf 1735, Madan.sreddy, Madman2001, Maethordaer, Magister Mathematicae, MagnaMopus, Magnoliasucks, Maharding, MaizeAndBlue86, Majorly, Mal4mac, Malleus Fatuorum, Malo, Manchurian candidate, Mancini141, Marek69, Mariraja2007, Mark.murphy, MarkTB, MarkThomas, Marmenta, MarsRover, Martial75, MartinDK, MartinHarper, Martinwilke1980, Martpol, Maruthi Achsara, Materialscientist, Matt Crypto, Matthewrbowker, Matticus78, Mattisse, Mattsarahcat, Mattu00, Maurice Carbonaro, Maurreen, Maximus Rex, Mbell, McGeddon, McLovin123459, McSly, Mclay1, Mehranwahid, Melongrower, Mentifisto, MetsFan76, Mgiganteus1, Michael Devore, Michael Hardy, Michael.goulais, Mifter, Mightymights, Mikael Häggström, Mike Peel, Mike Rosoft, Mike Storm, MikeBaharmast, Mindmatrix, Minesweeper, Minimac, MinorContributor, Miquonranger03, Mirv, Mjr162006, Mlhooten, Moingv, MongolWiki, Montrealais, Mookiefurr, Moomoomoostwos, Moosester, Mozerella8, Mr Adequate, Mr plant420, Mr.Kassner, Muffin123456789, Mufka, Muraabi6, Murgh, Myanw, Mygerardromance, Mzilikazi1939, N3bulous, NAHID, NHRHS2010, Nagualdesign, Napate, Naryathegreat, Nate5713, Nathan1991nathan1991, NativeForeigner, NatureA16, NawlinWiki, Nayvik, Nchaimov, Ncmvocalist, Nealmcb, Nehrams2020, Neilk9393, NellieBly, Nemu, Neural, Neurolysis, Neutrality, Neverquick, NewEnglandYankee, Newgrounder, Newone, NiallJones, Nickkid5, Nighthawk380, Nightscream, Nikhil Sasi, Nipponese, Nixdorf, Njaelkies Lea, Nk, Nlu, No Guru, Noctibus, Nothingisayisreal, Nseidm1, Nsmith 84, NuclearWarfare, Nuttycoconut, Octahedron80, Oda Mari, Ohnoitsjamie, OlEnglish, Old Moonraker, Oleg Alexandrov, Olliegodwin, Ollj, Olorin28, Olso12345, Omeed Was Here, Omicronpersei8, Onsly, Oraculo miraculoso, Orlady, Oscardove, OverlordQ, Oxymoron83, P123567890p, P3Pp3r, PFSLAKES1, Paidoantonio, Pakaran, Paranoid, Parresh, ParticleMan, Pascal666, Pass a Method, PaterMcFly, Patriarch, Patrick, Patssle, Patstuart, Pcorty, Peak, Pediainsight, Perfect Proposal, Permafry42, Perníček, Persian Poet Gal, Peruvianllama, Peter470, Peterdjones, Pharaoh of the Wizards, Phase Theory, Phil Boswell, Phil2324, Philip Trueman, Phlembowper99, Phobiaphobia, Picus viridis, Pinkadelica, Pizza Puzzle, Pluma, Plumbago, PoizonMyst, Pontoots9, Popol0707, Porthos0, Prancibaldfpants, PresN, PrestonH, Priyamd, Procatcher31, ProhibitOnions, Prunesqualer, Psy guy, Pumpmeup, Punkrockpiper, Purduecit, Pure maple sugar, Purnajitphukon, Pwnage09, Pyrophotographer, Pyrospirit, Python eggs, Q2op, Qazwsx197966, Quadell, Quantumor, Quintote, Quirkie, Qxz, R. fiend, RA0808, RG2, RJHall, Ramananrv123, Randolf Richardson, Random User 937494, RandomP, Randwicked, Rashaani, RazorICE, Razorflame, Rdht, Reaverdrop, Rebel700, Reidlophile, Remember the dot, Reneeholle, Renil, Retodon8, Rettetast, RevRagnarok, RexNL, Rgdboer, Rich Farmbrough, Rickmeister, Rickmeister6, Rillian, Rjwilmsi, Rmky87, Rnt20, Robbiemasters89, RobertMel, Roger Davies, Romanm, Rommopaula, Ronhjones, Rootswailer, Rossnixon, RotaryAce, Roy Brumback, Rrburke, Rror, Ruben3186, Rumping, Ruud Koot, Ryan, RyanParis, Ryuzakihateskira, S h i v a (Visnu), SDC, SFC9394, SJP, SMC, ST47, Saggipie, Sahsan, Sahuanimesh, Sajt, Sam Korn, Samoboow, Samuel, Sappho'd, Sardanaphalus, Saros136, SashaTheAwesome, Satori Son, Saturn star, Savantnavas, Sawadeekrap, Sbharris, Scepia, Sceptre, Schmackity, Scohoust, Scoo, Scopesmonkeys, Scottywong, Scroggie, Scwlong, Seaphoto, SeeYouNextTuesday999, Sengkang, SeniorInt, Serendipodous, Seresin, Setu, Shadowdemon936, Shadowjams, Shahab, Shai-kun, ShakataGaNai, Shalom Yechiel, Sharkface217, SheffieldSteel, ShinobiX200, Shirik, Shitonmydick1234, Shoeofdeath, Shreevatsa, Shroitman, Siddhant, Siddharth Prabhu, Sidoburn, Silly rabbit, Silvergoat, Sinep2, Sinteractive, Sir Nicholas de Mimsy-Porpington, Sjtarr, Skizzik, Skunkboy74, Sky83, Skyezx, Slightsmile, Slon02, Smallman21, Snoopdoug22, Snowolf, SoSaysChappy, Solphusion, Some jerk on the Internet, Someguyonthestreet, Someone0707, Sonikkua, Soriano9, SpaceFlight89, Spacegoat, Spacepotato, SparrowsWing, Spencer, Spig a digs, Splintercellguy, SpookyMulder, SpuriousQ, Squids and Chips, Squirtle1994, Srinivas, Sruk77, Srushe, Stannered, StaticGull, Steel, SteinbDJ, StephanKetz, Stephen G. Brown, Stephenb, Stevemanjones, Stevertigo, Stimulieconomy, Storm Rider, Strane1991, StrawberryPink, StuffOfInterest, Sudhir.routray, Suisui, Superbrainr, Superm401, Supraboy001, Sushant gupta, Synchronism, Synthiac, T g7, THEN WHO WAS PHONE?, TJSwoboda, Tapir Terrific, Tbhotch, Tcncv, TeaDrinker, TeleComNasSprVen, Telempe, Teles, Teles.ME, Tellyaddict, Tempodivalse, Terik brunson, Terper, Th1rt3en, The Bread, The Dark Side, The Evil Spartan, The High Fin Sperm Whale, The Rambling Man, The Thing That Should Not Be, The Transhumanist, The Wiki ghost, TheAlphaWolf, TheDJ, Thealltruth, Theda, Thedjatclubrock, Thematrix007, Theone70, Theoneintraining, Thewolrab, Thialfi, Thingg, Think outside the box, Three geeks, Thulasidharan2k8, ThuranX, Tide rolls, TimProof, TimothyRias, Timwi, TitanOne, Tnvkumar, Tobby72, Tom Kitt, TomasBat, Tommy2010, Tonezilla88, Tpbradbury, Tracy49, Trahern1994, Trec'hlid mitonet, Trevor Andersen, Trevor MacInnis, Treybien, Trilobitealive, Truthnlove, Ttony21, TutterMouse, Tvaughn05, Tylergriswold, Typuifre, UT200100, Ulflund, Ulric1313, Uncle Dick, Uncle G, Uncle Milty, UnitedStatesian, Uq, Uris, Utcursch, VBGFscJUn3, VLDG123, VQuakr, Valerius Myotis, Vanfug, Vanished User 0001, Vanished User 1004, Vanished User 4517, Vanished user 39948282, VanishedUser314159, Vedjain, Vedranf, Velella, Vendeka, Vercingetorix08, Vere scatman, Versus22, Vijinjain, VinnyXY, Vipinhari, Vishal144, Vishnava, Vitaliy skrynnik, Voortle, Vorpal blade, Vrenator, Vsmith, WadeSimMiser, Waggers, Wanze, Watch37264, Wavehunter, Wavelength, Wayne Hardman, Weatherman1126, Weirdo1990, WelcomeBackWinter, WelshMatt, West Brom 4ever, Whizmd, Who3, WikiSlasher, Wikiloop, WikipedianMarlith, Wikipelli, WikiworldJ, Wildnox, William Avery, WilliamKF, Williamrlinden, Willking1979, WillowW, Winegum74, Wintonian, Wireless Keyboard, Wirelesspp, Wjfox2005, Wlmg, Wmpearl, Woohookitty, Wooitscaroline, WookieInHeat, Woudloper, Wperdue, Wtmitchell, Wulfila, Wysprgr2005, X31forest, XJamRastafire, Xammer, Xeonxeon12, Xevi, Xiahou, Xiner, Xingzeng, Xlec, Xoxoxoxoxxx3, Xunknownxx, Xxuberzang, Xzuiko, YUL89YYZ, Yafujifide, Yamamoto Ichiro, Yddam, Yelyos, Yidisheryid, Yonatan, Yresh, Zachareth, Zamfi, Zazaban, Zeimusu, Zenithfel, Zeroparallax, Zfr, Zharradan.angelfire, Zkczkc, Zojj, Zoz, ביני35, علی ویی، قرابعنلعي کیب 2403 anonymous edits

**Polytheism** *Source*: http://en.wikipedia.org/w/index.php?title=Polytheism *Contributors*: 198.60.121.xxx, 207.93.56.xxx, 9258fahsflkh917fas, A8UDI, AA, Aarligbyrd, AdelaMae, Aethelmost, Agne27, Akbaralio, Akhilleus, Alansohn, Alborz Fallah, Amakuru, Aminullah, Andonic, Andre Engels, Andries, Andycjp, Angela, Angie Y., Antandrus, Antioco79, Antonio Lopez, Arb, Arjun01, Art LaPella, Aryaparis, AtZeuS, BD2412, Barbara Shack, Bart133, Beckster32, Ben Babcock, Ben Ben, Betterusername, Bhadani, Biebersbro3, Bindumadhava, Black and White, Bloodofox, Bobo192, Boing! said Zebedee, Brandon, Bremerenator, Bryan Derksen, Bryonmorrigan, Bubba hotep, Camw, CanadianLinuxUser, Cantiorix, Capricorn42, Catgut, Cesar27, CharlesMartel, Chemical Inbalance, Chris the speller, Chsbcgs, Cishaurim, Closedmouth, CommonsDelinker, Conjugations, Conversion script, Courcelles, Cubs197, Curps, D Monack, DAOUD9, DMacks, DNewhall, DVdm, Da Joe, DanielCD, Darklilac, Darth Panda, Davish Krail, Gold Five, Dbachmann, Deeptrivia, Delldot, Demetrius Zaarin, Deor, DerHexer, Dewrad, DjCarrad, Dougweller, Dpv, Dr. Günter Bechly, Dreadstar, Dream of Nyx, Dumcgr, DwightDBeasenbury, E. Fokker, EALacey, Edtheist, Eduardo Sellan III, Eequor, Ehrenkater, El C, ElKevbo, Eltigre007, EmilJ, Enok Walker, Enviroboy, Epbr123, Epoptes, EricSerge, Eumolpo, FT2, Falcon8765, Fang 23, Fastily, Feureau, Fitzgiggler, Flo98, Frankenpuppy, Freakofnurture, Fubar Obfusco, Fullstop, Gaius Cornelius, Gakusha, Gary King, GhostPirate, Giannisf, Glenn, Goethean, GorillaWarfare, Grammar conquistador, Gregbard, Gregzeng, Gulmammad, Gwendal, Habibko, Hadjalexiou, Haham hanuka, Hamster X, Haukurth, HenryLi, Hokninja, Hu, Hydrogen Iodide, Héliocoptero, I2E4S6, IW.HG, Ian Pitchford, Ig0774, Ilkali, InFairness, Infrogmation, InverseHypercube, Ionescuac, Ism schism, Izakalka, Izehar, J. W. Love, J.delanoy, JForget, JHunterJ, Jackfork, JahaalChakravarti, JamesBWatson, Jauerback, Jauhienij, Jayjg, Jeandré du Toit, Jeff3000, Jeffrey O. Gustafson, Jeffro77, Jesucristo301, Jguk, Jimmilu, John Cho, John254, Jorfer, Jossi, Julian Mendez, K7aay, Kahasabha, Kanook, Kathryn NicDhàna, Keegan, Keraunos, Kingfish, Kitty555, Kittymonk4, Kkrystian, Krsont, L Kensington, LAX, Larry_Sanger, Lastresort9292, Laterensis, Le Anh-Huy, LeaveSleaves, Leolaursen, Lexor, Liftarn, Lincmad, Lolcatas, LordSimonofShropshire, Lowellian, Luna Santin, Lyonspen, MER-C, MWShort, Machine Elf 1735, Madhero88, Magicalsaumy, Manu bcn, Martha Helene Jones, Martin451, Maximus Rex, Mdiamante, Megan1967, Meic Crahart, Mel Etitis, Meno25, Merope, Mezigue, Michael Hardy, Michael%Sappir, MichaelMaggs, Midnightblueowl, Mike Crahart, Mild Bill Hiccup, Miquonranger03, Miss Bea Haiving, Mlessard, Monkeykiss, MorganaFiolett, Mueka238, Munci, Mygerardromance, NantonosAedui, NeilN, Neurolysis, Nikai, Nixeagle, Nneonneo, Nsaa, Oblivious, Oge Naws, Ojigiri, Onorem, Oolong, Ospalh, Pacific PanDeist, Paine Ellsworth, PartTimeGnome, Paul August, Peter Kirby, Pigman, Platonic Guardian, Plumcherry, Prodego, Pujo1, Pundit, QaBobAllah, R'n'B, Raj2004, Rakista, Random articles, Renaissancee, Rhobite, Rich Farmbrough, Richard Keatinge, Rickyrab, Rise Above the Vile, Rjwilmsi, Roadrunner, Rossen3, Rsrikanth05, RunOrDie, SFH, SWAdair, Sam Korn, SchfiftyThree, Seidkona, Septegram, Shii, Shirik, Shlomke, Skapur, Snowdog, So God created Manchester, SpaceFrog, Sparklenbling, SteinbDJ, Stevepeterson, Stormerne, Susan Mason, SwissCelt, TakaraLioness, Tckma, Teardrop onthefire, The Cunctator, The One True Fred, The Transhumanist, The sunder king, Thesloth, Thingg, Tide rolls, TimBentley, Timir2, ToddWC, Todgar, Tricolor Truce, Trusilver, Tstrobaugh, Umbrellas, Uncle Dick, Useight, UtherSRG, V, Vaka1258, VanishedUser314159, VedicScience, Viscious81, Vrenator, Vssun, WarFox, Wayne Slam, Wesley, Wetman, Whateley23, Wikislemur, Wisco, Wknight94, Wtmitchell, XJamRastafire, Xuchilbara, Yamamoto Ichiro, YouMoo!, Yworo, Z054J, ZepperZepper, Zookeep, 703 anonymous edits

**Paganism** *Source*: http://en.wikipedia.org/w/index.php?title=Paganism *Contributors*: 100110100, 24.251.118.xxx, 999, A Macedonian, ALR, Abb615, Abdul Muhib, Abdullahazzam, Abdullais4u, Abm360, Achriel, Ackie00, Activistrep, AdelaMae, AdjustShift, Aetheling1125, Aka042, Aldie, Aldnonymous, Aleister Wilson, AlexTiefling, AliaGemma, Alkados, Allgood1, Amcaja, Americist, Amikeco, AnakngAraw, Andries, Andycjp, Anglius, AnonMoos, Ante Aikio, Anti-pagan, Apollo, Aquilina, Arthur Warrington Thomas, Ascánder, Ashdurbat, Ashley Y, Ashleyy osaurus, Asnac, Attafei, Audiosmurf, Auric, Auréola, Avellano, Avoided, Awakeandalive1, B0Rn2bL8, BD2412, Badagnani, Barista, Before My Ken, Begoon, Ben D., Ben-w, Benhealy, Berig, Bharatka, Big Adamsky, BillWSmithJr, Billionpants, Billscottbob, BirgitteSB, Blessedbe 13, Bloodofox, Bloodshedder, Blueboar, BobMak, Bobo192, Bodnotbod, Bongwarrior, Borock, Borovi4ok, Bowmasterkratos, Bp13152251351420, Braapar2, Bradeos Graphon, Bradv, Brandmeister (old), Brett.donald, Brian Crawford, Brighterorange, Bubbadrew2, BurnDownBabylon, CALR, Caiaffa, Caltas, CambridgeBayWeather, Can't sleep, clown will eat me, Capike, Cardreader, Carl.bunderson, Carlaude, Catinthehat93, Cavalorn, Cbaer, Cbdorsett, Ceaser, Celtus, Ceyockey, Charles Matthews, Charred Feathers, Chasclifton, Ched Davis, Chrislk02, Christopher140691, Coelacan, Conversion script, Courcelles, Coveritus, Cremepuff222, Crzer07, Cyde, DO'Neil, DVD R W, DVdm, DakotaWild, Damac, DanMS, DanielCD, Danski14, Darthanakin, Daven200520, Dawn Bard, Dbachmann, Dchmelik, DeadEyeArrow, Deepak.mr, Dekisugi, Delldot, Demoskinos, Demosthenes426, Denihilonihil, Dennis Valeev, DerHexer, Derek Ross, Dethme0w, Deviantsoul67, Dfrg.msc, Dirkbb, Discospinster, Dmerrill, DocWatson42, Dogface, Don Braffitt, Donmac, Dougweller, Dpv, DrDnar, DrNumLock, Drax10, Dreadstar, Dream of Nyx, DreamGuy, Dysepsion, ESkog, Eanudimmud, Earth, Ed g2s, Edcolins, Editor2020, Edmonton goddess, Eequor, Eih066, El C, Eliz81, Erolos, Esimal, Estéban, Eulalie Écho, Eupedia, Everyking, Evmore, ExRat, Exxolon, Faradayplank, FisherQueen, Formicarius, Foxfollower, Fraggle81, Francs2000, Fratrep, Frau Drache, Frehley, Frosted14, FunPika, FuroraCeltica, Fuu(k face, Fuzzypeg, G.W., GNSBraveheart, Garik, Garjják, Gdje je nestala duša svijeta, GideonF, Gilliam, Gitana7, Gjs238, Glenn, Godsgroupie, Gogo Dodo, GoldenMeadows, Golgofrinchian, GraemeL, Graham87, Greenshinobi, Gregchapman, Gregums, Gryffindor, Gunnar Hendrich, Haiduc, HalfShadow, Hao2lian, Harro5, Haukurth, Hbdragon88, HeartofaDog, Heron, HexaChord, Highpriority, Hilvert, Hippalus, History2007, Hroðulf, Hu, IAF, IamthatIam, Ian Pitchford, Ian.thomson, Iasos, Ihcoyc, Inspector 34, Intr1400cc, Iridescent, IronMaidenRocks, Ish ishwar, J.delanoy, JNW, Ja 62, Jag123, Jamon8888, January, Japanese Searobin, Jason Farrow, Jebus989, Jefferson Anderson, Jensbn, Jesus Stobie, Jguk, Jkelly, Jken89, Jni, JohnArmagh, JohnCD, Johntcullen, Jonathunder, Jonpro, Joseph Solis in Australia, Josiah Rowe, Jtnelson, Junjk, Justin Eiler, K7aay, Kaiser matias, Karol Langner, Karonaway, Kathryn NicDhàna, Kcajsitruc, KcinSometin, Keefygee, Keepssouth, KelsoLovesJesus, Kim Dent-Brown, Kinu, Klafubra, KnowledgeOfSelf, Koavf, Konigsteiger, Krawi, Krukouski, Kukini, Kyle Phoenix, La goutte de pluie, LachlanA, Lacrimosus, Lammidhania, Latulla, Laurel Bush, LaurenKaplow, Lectiodifficilior, Leithp, Lendorien, LiftWaffen, Liftarn, Lightmouse, Lihaas, LinasLit, Lolwut0-0, Looper5920, Lorynote, LotR, Lotje, Lottiotta, Lucyintheskywithdada, Luke chan91, Lullabyluna, Lysdexia, MPF, MSJapan, Macedonian, Machine Elf 1735, Madchester, Made of nguyen, Magister Mathematicae, Maiko90, Majorly, Malvikabaru, Man vyi, ManOnPipes, Marcussantiago53, Marin4usaka, Materialscientist, Mattis, Maximaximax, McSly, Mceder, Melamed katz, Miaow Miaow, Michael Hardy, Midnightblueowl, Mike Rosoft, Mikerock, Mild Bill Hiccup, Million Moments, Millionpants, Milwaukee3181, Minimac, Mirwin, MishaPan, Mister X, Modulatum, MoiraAshleigh, Montgomery '39, Moonstar6, Morningstar2651, Mow742369, Mr. Absurd, Mr. XYZ, MrMizfit, Mushroom, Muski27, Mwanner, Mwilso24, Mythdon, Mzjohnna, N1RK4UDSK714, Nakon, NatusRoma, Neelix, Netscott, Neutral current, Ninly, Nio-guardian, Nirvana2013,

Nonexistant User, Nymf, OddAud, Odin84, Odinbolt, Ohnoitsjamie, Olaf Davis, Olessi, Oliver Lineham, Olivier, OllieFury, Omicronpersei8, Omnipaedista, Onaga, One-dimensional Tangent, Orangemarlin, Oteta, Owen, P.D., Padan, Pagan Witch90, Pahari Sahib, Panairjdde, Patch666, Paul August, Paul-L, PaulGregory, Pb30, Pde, PelleSmith, Pennygains, PeterBeckley, Philipholden, PhnomPencil, Picapica, Pietrow, Pigman, Pilotguy, Pollinator, Pouya, Preachur, Priestess Jean, Protonk, Qmwne235, Quadell, Quintote, R'n'B, RJPe, Raven in Orbit, Rawling, Rds865, Rebecca2008, RedWolf, Redfarmer, Redheylin, Redtigerxyz, RevMarilyn, Rich Farmbrough, Rising*From*Ashes, Rivertorch, Rjwilmsi, Rmosler2100, Robert K S, RobinHood70, Romaioi, Ronhjones, Rossen4, RoyBoy, Rucha58, SF007, SUL, SWAdair, Saga City, SaintHammett, Sam Spade, Samtar, Sannse, Santamoly, SarekOfVulcan, Satanael, Scarian, Scissor sisters123, Sdornan, Sehnga, Seidkona, Sessee, Seth ze, Shaddack, Shan Jayran, Shanes, Shantrika, ShelfSkewed, Shii, Shivian, Siddownwaldo, Sk8rhawk, Skamecrazy123, Skater3911, Slicing, Slightsmile, Smalljim, SoCalLisa, Some jerk on the Internet, Someone65, Sophie means wisdom, SpaceFlight89, SpaceLem, Spliffy, SpookyMulder, SpuriousQ, Srnec, StAnselm, Star667, Starfoxy, Stbalbach, Stefania wiki, Stevenmitchell, Stevertigo, Sunray, Sword of Light, T3gah, TAnnison, TDogg310, TKD, Taam, Tamfang, The Wednesday Island, The lost years, The-Watermammal, TheMadBaron, TheNewPhobia, Thermionic1, Thingg, Thorskegga, Tide rolls, Tim Long, Tisquantum, Toadsboon, Toddsschneider, Tomeasy, Toscaesque, Touch Of Light, Transity, Tregoweth, Treisijs, Triskaideka, Trjn, TrustTruth, Twinsday, Twsx, UNIT A4B1, Ultrogothe, Unchqua, Unixer, Unless 68, Untifler, Urzumph, Userafw, Uther67, UtherSRG, Vaderalex, Vaughanwj, Vegetator, Viande hachée, Vianello, VirtualDelight, Volrath50, Vrenator, W.andrea, Walton One, Wandering Writer, WarriorPrincessDanu, Wavelength, WeniWidiWiki, Wetman, Whiteash, Wiglaf, Wikiacc, Wikiality123, Wikignome0530, William915, Wilson44691, Wimt, Winertai, Woohookitty, Wyattb995, XLR8TION, Xhaoz, XmetalX, XtraordinaryX, Xxglennxx, Yamaka122, Yamamoto Ichiro, Yanamad, YellowMonkey, Yerpo, Yidisheryid, Zachary4959, Zachbe, Zandperl, Zatch908, Zenben13, Zenswashbuckler, Ziusudra, Zoeyeve, ZondaSX, Zoohouse, 1087 anonymous edits

**Pantheism** *Source*: http://en.wikipedia.org/w/index.php?title=Pantheism *Contributors*: .snoopy., 1ForTheMoney, 1exec1, 2D, 613kpiggy, AED, Aaron McDaid, Abhishek191288, Accurrent, Acidmonkey, Acmartin, Adamahill, Afghana, Ahoerstemeier, Akaqueequeg, Aknxy, Algont, Alphakimori, Alphathon, Altenmann, Amalas, Ameliahupe, Andres, Andrew Filipe, AnnaFrance, Antandrus, Aorbar, Aquillion, Arb, Arda Xi, Aregakn, Arker, Armyrifle9, Arny, Artichoker, AwkwardSocks, Axeloide, Aymansousa, B9 hummingbird hovering, BD2412, BM, Baa, Babakathy, Barticus88, Bcorr, Beno1000, Bkonrad, Bkwillwm, Bomac, Bowei Huang 2, Brian0918, Brighterorange, Brsullivan, Bryan Derksen, BullRangifer, CRGreathouse, Caleb Murdock, Calton, CambridgeBayWeather, CanadianLinuxUser, CanisRufus, CardinalDan, Carlo.Ierna, CharlotteWebb, ChazMercer, Chgwheeler, Chsbcgs, Clerks, Closedmouth, Colonies Chris, Conversion script, Coolian, Cpiral, DDerby, DT Strain, DVD R W, Dainis, Dalejsprague, Damian James, Dave314159, David Gerard, David Ludwig, Deconstructhis, Deepblackwater, DeistCosmos, Deklund, Dennis Brown, DevOhm, Dinkytown, Dlohcierekim's sock, Dnagod, DoomsDay349, Dravida Dasa, Drbreznjev, Dustinasby, Dysprosia, ERcheck, Edward321, Eequor, Efguerre, El C, Epsilon60198, Erics, Exayda, Exult, FayssalF, Feedmecereal, Fieldday-sunday, Finbarr Saunders, Firsfron, Flakture, Flange the Flee, Flyingp, Fratrep, Fuzheado, Fuzzypeg, Gadfium, Gaius Cornelius, Galileo13, Galileo313, GirasoleDE, Glane23, Glenn, Goethean, Gogobera, Gr8opinionater, Grafen, GraysonGraham, Gregbard, Greier, Guslarkachic, Hadal, Hammaad, Hasam, Headbomb, HelioSmith, Hendrixx01, Hydrargyrum, I'm so Kewall, II MusLiM HyBRiD II, IW.HG, Igiffin, Igni, Ikana, Imc, Infrogmation, Interwal, InverseHypercube, Isnow, Isospin, Itain'tsobad, J.delanoy, JASpencer, JWSchmidt, JakeVortex, Jamesgibbon, Janet Davis, Java7837, Jdhunt, Jeff3000, Jeff5102, Jenn7723, Jfdwolff, Jlrobertson, John Bessa, John D. Croft, JohnDelano, JonHarder, JonSangster, Jorfer, Jossi, Joyous!, Jrun, Judeeylander, Juliancolton, Jwrosenzweig, Jóna Þórunn, K kisses, Kafziel, Kavita Kapoor, Kenneth Cooke, Keraunos, Keresaspa, Kimania, Kimiko, Kingturtle, KittyKAY4, Kku, Klaxon, Koavf, Kokoriko, Kpjas, Kross, Kzollman, L Kensington, Lawrence King, Lefty, Lemontree22, Lestrade, Liftarn, LikeHolyWater, LindsayH, Looxix, LordSimonofShropshire, Lotje, Lst27, M.Imran, MAurelius, MGTom, Machine Elf 1735, Mackan79, Magister Mathematicae, Maheshkumaryadav, Malik Shabazz, Malik047, Marbehraglaim, Martin Berka, MattOates, Max314, Maximus Rex, Mel Etitis, Michael Hardy, MikeCapone, MikeEagling, Mikenassau, Milad10us1985, Mild Bill Hiccup, MiracleMat, Mirandansa, Mirv, Modernist, Momoricks, Moreschi, Mourad EL AMRANI, Mr.98, Mushroom, Mystic.Ventus, NCDane, NatDemUK, Natural84, Naturalistic, Naturyl, NawlinWiki, Neelmack, Nehrams2020, NickelShoe, Noosphere, NuclearWarfare, Nunquam Dormio, Nyenyec, OakMt, Omnipaedista, Optim, Owen, Pacific PanDeist, Paine Ellsworth, Pandeism debate, Peruvianllama, Phil Bridger, Phoenixrod, Physicistjedi, PiMaster3, Pigman, PinchasC, Pioneer-12, Platonic Guardian, Pollinosisss, Poppafuze, ProductionsGuy, Purple Paint, Qwertyus, RDF, Rachaelbyrnes, Raj2004, Rakista, RedWolf, Redheylin, Ren Sydrick, Renesis, Requestion, Rewinn, Reyk, Richard001, RickK, Rjwilmsi, Robert Brockway, RonDivine, Roy Brumback, RyanCross, SGreen, SKEpptic, Sam Spade, Samwaltz, Sangitalaya, Saturnight, SchfiftyThree, Schmiteye, Scott Burley, Seventhcroak, Shibboleth, Shythylacine, SimonP, Singinglemon, Sketch051, Skomorokh, Slrubenstein, Snowgrouse, Snoyes, Some jerk on the Internet, Sophus Bie, SteinbDJ, Syncategoremata, T3h 1337 b0y, Taram, Tarotcards, Tedbradford, Texture, The Egyptian Liberal, The Random Editor, The Sanity Inspector, The Thing That Should Not Be, Themfromspace, Thereisstillhopeinone, Thirdeye4u, This user has left wikipedia, Timwi, Tomalak geretkal, Topilsky, Torquemama007, Tothebarricades.tk, Tracker1312, Twas Now, Twsx, Uncle Dick, Unimath, Unmerklich, UtherSRG, Vacuum, Vaikunda Raja, Varttik, Vortexengineer, Wassermann, Wavelength, Welsh, Wetman, Whatyoudontsee, White Shadows, Wik, Wikipediaphile, Wikivedalankara, Wikiwikiwoo2, Will Beback, Will Beback Auto, WillMall, Wordock2, Worldpan, Yintan, ZZyXx, Zeupater, Zsero, Zvis, محمد, 805 anonymous edits

**Animism** *Source*: http://en.wikipedia.org/w/index.php?title=Animism *Contributors*: 3 Eyed Pete, 65.68.87.xxx, ABF, ACSE, Abc518, Abeg92, Aditya Kabir, Adjohnson916, Aetersyl, Ahoerstemeier, Aiken drum, Akira-otomo, Akuyume, Ala.urbane, Alan Liefting, Alansohn, AlotToLearn, Alrick, Alsandro, Ama1992, Amakuru, Andre Engels, Andrea105, Andres, Andycjp, Antandrus, Anubis Godfather, Asarelah, Asibikaashi, Astillac, Augen Zu, AzaToth, B, B., BD2412, Badger151, Bdean1963, Belovedfreak, Ben Babcock, Beno1000, Bepp, Berkut, Binary TSO, Black and White, Bluemoose, Bobo192, Bped1985, Brainyiscool, Brendanq77, Brianski, BrotherGeorge, Bryan Derksen, Btj, Bushlala, CJLippert, CRGreathouse, Cacala17, Cacycle, CambridgeBayWeather, Canderson7, Caniago, Cannibalfox, CardinalDan, Catinthehat93, Charles Matthews, CiTrusD, Ciup, Closedmouth, Cocoaguy, CommonsDelinker, Computerjoe, Conversion script, Cool3, Counterfact, CrossHouses, Csquad27, Cwmhiraeth, DBlomgren, DVD R W, Da Joe, Daniel11, Daniel73480, Darth Panda, Dbachmann, Dbenbenn, Dcoetzee, DeLarge, Debresser, DevOhm, Dillard421, Discospinster, Distal24, Donatus, Dream of Nyx, Drewalanwalker, Drmies, Dwmr, Dycedarg, Dysprosia, ESkog, Editor2020, Edward321, Eequor, Ehn, El C, Elcalen, Ent, EoGuy, Epbr123, Epsilon60198, Equine-man, Erget2005, Estudiarme, FF2010, Fang 23, Fastily, Feeeshboy, Fireblade21, Fishwristwatch, Foant, Frankenpuppy, Freakofnurture, Fuddle, Gary King, Geeoharee, Gilliam, Glen, Glenn, Goldilox369, Graham87, Gregbard, Gregory Benoit, Gurubrahma, Guslarkachic, Hajor, Harryboyles, Hashar, Heike B., Heimstern, Helga76, Heligan, Hmains, Homagetocatalonia, Hu, Hurricane111, Ian Spackman, Icairns, Iliketomeow, ImperatorExercitus, Inge-Lyubov, Ishmaelblues, Jan1nad, Jarble, Jason.stover, Jayjg, Jcvamp, Jeff G., Jeff3000, Jennavecia, Jim Douglas, Jkelly, Jmabel, Jmrowland, Jncraton, JoJan, John D. Croft, Johnboyloy, Jonathan.s.kt, Jrockley, Jrtayloriv, Jtkiefer, JuJube, Jumbolino, Justanother, Justinodem, K kisses, Kablammo, Kalogeropoulos, Kathryn NicDhàna, Kdau, Kelovy, Kesac, Kingturtle, Kirachinmoku, Klnosaj, KoobieKitten, Kowey, Kroose, Kross, Kummi, Kuru, Kutera Genesis, Kwamikagami, L Hamm, LFaraone, LWG, LadyofShalott, Lambiam, Leadwind, Leafyplant, LeilaniLad, Leks81, Leuko, Lev lafayette, Liftarn, LilHelpa, LittleHow, Lololapplejacks, Louiebb, Lovecz, Lucyintheskywithdada, M1k3y, MA (Cantab), MGTom, Machine Elf 1735, Magioladitis, Maksym Ye., Malik047, Marek69, Materialscientist, Matthew Auger, Maziotis, Meaghan, Mensch, Mentifisto, Merovingian, Meursault2004, Michael Hardy, Midnightblueowl, Milkyface, Millennium Twain, Mindstalk, Miquonranger03, Mix Bouda-Lycaon, Mkjo, Montrealais, Moomuu, Moondyne, Mporch, MrKeith2317, Musoniki, Myanw, Naniwako, Naohiro19, Natalie Erin, NatureA16, Naturyl, Neelmack, NellieBly, Neokamek, Neutrality, Neverquick, Ngouandi, NickelShoe, Nihil novi, NorwegianBlue, Number 0, Numenetics, Objecthood, Odinsjarl, Onco p53, Orz, PTSE, Packers19960, Pascal.Tesson, Pdinc, Pearle, Pedro, PelleSmith, Pernambuco, Pgdudda, Phatius McBluff, PhilKnight, Philip Trueman, Pigman, Plesiosaur, Politepunk, Possum, Primate Sapiens, Procrastinatrix, Prodego, Ptcamn, Puchiko, Punarbhava, Pwqn, QASIMARA, QVanillaQ, Qgil, Quebec99, Quinsareth, Quintote, Qxz, RPellessier, RadioFan2 (usurped), Ralf 58, Ramaksoud2000, Reedy, Rettetast, Rhankins, Rholton, Rhysjsk, Rich Farmbrough, Rjwilmsi, Romanm, Ronhjones, Rschen7754, Ruud Koot, Saga City, Sam Spade, Sanawon, Sander123, ScAvenger lv, Scarian, Schwael, Scoobysam100, Sebesta, Shadowlink1014, Shanoman, ShiMutale, Shii, Shirulashem, Sitush, Skp2y F thorax, Slrubenstein, Sluzzelin, Slysplace, Snarius, Snowolf, SoWhy, Sofeil, SonOfNothing, Special-T, Spongefrog, Steel, Stephen Bain, SteveRamone, Stevey7788, Sting au, Stricklywiki, Stwalkerster, Suske422, Syrthiss, TUF-KAT, Tamsier, Texture, The Thing That Should Not Be, The Transhumanist, The Ungovernable Force, The-Pope, TheMadBaron, TheTrueSora, Thismightbezach, Tiddly Tom, Tiptoety, Titanium Dragon, Tjbird9675, Tommy2010, Topilsky, Trialsanderrors, Trusilver, Ttwaring, Turian, TutterMouse, Ugur Basak, Ulric1313, Unitedstates1000, VKokielov, Vald, Vardion, Vegaswikian, Velella, Versageek, Vicki Rosenzweig, Viriditas, Vishnu2011, Wackywace, Waggers, WereSpielChequers, Wichienmaat, Wiki alf, Wiki libs, Wiki-uk, Wolfdog, Woohookitty, X!, YankeeInCA, Yettie0711, Yst, Zachorious, ZapThunderstrike, Zcratsley, Zzuuzz, 812 anonymous edits

**Natural World (TV series)** *Source*: http://en.wikipedia.org/w/index.php?title=Natural_World_%28TV_series%29 *Contributors*: Alexgray, Baguala, BaldBoris, Betacommand, Boothferry, David Newton, F W Nietzsche, Hibernian, IanC, Josh of the Decade, Josiah Rowe, Mattbr, Mepolypse, Mhiji, Mspraveen, Msrasnw, Otto4711, Perodicticus, PhilKnight, Pigsonthewing, Steinsky, The JPS, WadeSimMiser, Woohookitty, 13 anonymous edits

**Reverence** *Source*: http://en.wikipedia.org/w/index.php?title=Reverence *Contributors*: Ary29, Ffgamera, JD554, JHunterJ, John Depp, Lankiveil, Metalindustrien, NawlinWiki, Neelix, Penguinwithin, R'n'B, Rjakew, Thecheesykid, Wolfpeaceful, Xyzzyplugh, 8 anonymous edits

**Tree worship** *Source*: http://en.wikipedia.org/w/index.php?title=Tree_worship *Contributors*: Aiken drum, Andycjp, Aodhdubh, AshLin, Bloodofox, Curtis Clark, Dbachmann, Deanlaw, Faolin42, Inwind, Jccooper, John S Moore, JubalHarshaw, MKleid, Mairi, Maustrauser, Mellamoandres, Metallurgist, Neelmack, Pearle, Pepertje, Pigman, Randommelon, Sival1979, Snori, Superk1a, Talking.kerbau, Trulystand700, WDavis1911, Wetman, Wiki-uk, Wingedsubmariner, Woohookitty, 26 anonymous edits

**Animal worship** *Source*: http://en.wikipedia.org/w/index.php?title=Animal_worship *Contributors*: ***Ria777, -Ril-, 40fifw0, ACSE, Aaron Solomon Adelman, AgnosticPreachersKid, Altaïr, Andrea105, Andries, Andycjp, Angusmclellan, Anwar saadat, Apeloverage, ArielGold, Artefactual, Arun K Sivanandan, Ashley Y, BD2412, BMF81, Bearcat, Bebestbe, Beland, Brz7, Bsod2, Bueller 007, Buzancar, Cynwolfe, D-Rock, Dante Alighieri, Darklilac, Dbachmann, Dimitrii, Drmies, Ebear422, Elf, Everard Proudfoot, EvgenyGenkin, Fang 23, Fayenatic london, Gaius Cornelius, GimliDotNet, Hardyplants, Hibernian, HopeSeekr of xMule, Hottentot, Hvn0413, Ian.thomson, Invertzoo, Issuesixty soulsgreat, J04n, JaGa, Jossi, Kai Ojima, Katharineamy, Kraftiga, Legeran, Leo1410, M-le-mot-dit, MarcoLittel, Martinevans123, Materialscientist, Meerkatakreem, Megan1967, Musty man 01@hotmail.com, Mysticfeline, Natalie Erin, Neelmack, Nefertum17, Nehrams2020, Nick Number, Nmhall, PM800, PaleZoe, Papergirl10, Paul August, PelleSmith, Pigman, Pointillist, Pratheepps, Ptdtch, Punarbhava, Pwqn, RedWordSmith, Redtigerxyz, Ruud Koot, Salix alba, SamuelTheGhost, Seb az86556, ShelfSkewed, SiGarb, Sigurd Dragon Slayer, Simian monk, SlimVirgin, Sluzzelin, Spacexplosion, Steinsky, Stemonitis, Stephenb, T-borg, That Guy, From That Show!, Tizio, Tommy2010, Topbanana, Torchwoodtwo, Tripping Nambiar, Tydaj, Ulric1313, Varlaam, Velgean, Vipinhari, Viriditas, WereSpielChequers, Wiki-uk, William Ortiz, Woohookitty, WookieInHeat, Xezbeth, Xod, Yamaguchi, Zalpacuiztli, Zundark, 223 anonymous edits

**Astrolatry** *Source*: http://en.wikipedia.org/w/index.php?title=Astrolatry *Contributors*: 5 albert square, BD2412, Beeblebrox, Bigstroonz, Chironfletcher1212, Dbachmann, Dream of Nyx, Evilgohan2, Gaius Cornelius, Guslarkachic, Ian.thomson, Icairns, Kesaloma, Lotygolas Ozols, SummerWithMorons, Svanslyck, Torquemama007, Twilight Aruseus, Wiki-uk, Wiqi55, 14 anonymous edits

**Sacred mountains** *Source*: http://en.wikipedia.org/w/index.php?title=Sacred_mountains *Contributors*: Abductive, Auntof6, BD2412, Backeby, Beezhive, Celique, Dbachmann, Eugene-elgato, Fransvannes, Grp19, Iridescent, LilHelpa, M-le-mot-dit, Neelmack, PigFlu Oink, Piotrus, Princessofinfinitespace, R'n'B, Shinju, Shinkansen Fan, Woohookitty, Xufanc, Zozo2kx, 37 anonymous edits

# Image Sources, Licenses and Contributors

**File:Hopetoun falls.jpg** *Source*: http://en.wikipedia.org/w/index.php?title=File:Hopetoun_falls.jpg *License*: unknown *Contributors*: Diliff

**File:Bachalpseeflowers.jpg** *Source*: http://en.wikipedia.org/w/index.php?title=File:Bachalpseeflowers.jpg *License*: unknown *Contributors*: User:ZachT

**File:Galunggung.jpg** *Source*: http://en.wikipedia.org/w/index.php?title=File:Galunggung.jpg *License*: unknown *Contributors*: R. Hadian, U.S. Geological Survey

**File:The Earth seen from Apollo 17.jpg** *Source*: http://en.wikipedia.org/w/index.php?title=File:The_Earth_seen_from_Apollo_17.jpg *License*: unknown *Contributors*: NASA. Photo taken by either Harrison Schmitt or Ron Evans (of the Apollo 17 crew).

**File:Tectonic plate boundaries.png** *Source*: http://en.wikipedia.org/w/index.php?title=File:Tectonic_plate_boundaries.png *License*: unknown *Contributors*: Jose F. Vigil. USGS

**File:Hyperia.jpg** *Source*: http://en.wikipedia.org/w/index.php?title=File:Hyperia.jpg *License*: unknown *Contributors*: Chris.urs-o, Dodo, E-boy, John feather, Lycaon, Shyam, 1 anonymous edits

**File:Pangea animation 03.gif** *Source*: http://en.wikipedia.org/w/index.php?title=File:Pangea_animation_03.gif *License*: unknown *Contributors*: Original upload by

**File:Lightnings sequence 2 animation.gif** *Source*: http://en.wikipedia.org/w/index.php?title=File:Lightnings_sequence_2_animation.gif *License*: unknown *Contributors*: User:Koba-chan

**File:Top of Atmosphere.jpg** *Source*: http://en.wikipedia.org/w/index.php?title=File:Top_of_Atmosphere.jpg *License*: unknown *Contributors*: NASA Earth Observatory

**File:Dszpics1.jpg** *Source*: http://en.wikipedia.org/w/index.php?title=File:Dszpics1.jpg *License*: unknown *Contributors*: Daphne Zaras. Original uploader was Runningonbrains at en.wikipedia

**File:44 - Iguazu - Décembre 2007.jpg** *Source*: http://en.wikipedia.org/w/index.php?title=File:44_-_Iguazu_-_Décembre_2007.jpg *License*: unknown *Contributors*: User:S23678

**File:Ocean from Leblon.jpg** *Source*: http://en.wikipedia.org/w/index.php?title=File:Ocean_from_Leblon.jpg *License*: unknown *Contributors*: Amarvudol, CarolSpears, Erico Tachizawa, Zephynelsson Von, Zimbres

**File:Lake mapourika NZ.jpeg** *Source*: http://en.wikipedia.org/w/index.php?title=File:Lake_mapourika_NZ.jpeg *License*: unknown *Contributors*: Richard Palmer

**File:Mill Pond Sunset.jpg** *Source*: http://en.wikipedia.org/w/index.php?title=File:Mill_Pond_Sunset.jpg *License*: unknown *Contributors*: Eriky8 (Kevin Johnson)

**File:View from Cairo Tower 31march2007.jpg** *Source*: http://en.wikipedia.org/w/index.php?title=File:View_from_Cairo_Tower_31march2007.jpg *License*: unknown *Contributors*: User:Arad

**File:Hawaii Creek.jpg** *Source*: http://en.wikipedia.org/w/index.php?title=File:Hawaii_Creek.jpg *License*: unknown *Contributors*: God of War

**File:View of loch lomond.JPG** *Source*: http://en.wikipedia.org/w/index.php?title=File:View_of_loch_lomond.JPG *License*: unknown *Contributors*: Abubakr Hussain, Mohammed-Hayat Ashrafi, Maaz Farooq, Farmaan Akhtar, Mohammed Shah

**File:Chicago Downtown Aerial View.jpg** *Source*: http://en.wikipedia.org/w/index.php?title=File:Chicago_Downtown_Aerial_View.jpg *License*: unknown *Contributors*: Adrignola, Conscious, DavidLevinson, JeremyA, Kri, Leslie, Matt314, Quasipalm, 1 anonymous edits

**File:Biogradska suma.jpg** *Source*: http://en.wikipedia.org/w/index.php?title=File:Biogradska_suma.jpg *License*: unknown *Contributors*: User:Tsnena

**File:Malards in Golden Gate Park.jpg** *Source*: http://en.wikipedia.org/w/index.php?title=File:Malards_in_Golden_Gate_Park.jpg *License*: unknown *Contributors*: User:Mbz1

**File:Amazon Manaus forest.jpg** *Source*: http://en.wikipedia.org/w/index.php?title=File:Amazon_Manaus_forest.jpg *License*: unknown *Contributors*: Phil P Harris.

**File:Yellow mite (Tydeidae) Lorryia formosa 2 edit.jpg** *Source*: http://en.wikipedia.org/w/index.php?title=File:Yellow_mite_(Tydeidae)_Lorryia_formosa_2_edit.jpg *License*: unknown *Contributors*: User:Fir0002, user:brian0918

**File:Diversity of plants image version 5.png** *Source*: http://en.wikipedia.org/w/index.php?title=File:Diversity_of_plants_image_version_5.png *License*: unknown *Contributors*: User:Rkitko

**File:Animal diversity.png** *Source*: http://en.wikipedia.org/w/index.php?title=File:Animal_diversity.png *License*: unknown *Contributors*: User:Anilocra, User:Bkmiles, User:Kevincollins123, User:Lviatour, User:Medeis, User:Nhobgood, User:Opoterser, User:Panda3, user:Australianplankton

**File:Na Pali Coast - Kauai.jpg** *Source*: http://en.wikipedia.org/w/index.php?title=File:Na_Pali_Coast_-_Kauai.jpg *License*: unknown *Contributors*: Diaa abdelmoneim, Dmason4399, Hbk33

**File:Dendrarium Sochi Mauritanian arbour.jpg** *Source*: http://en.wikipedia.org/w/index.php?title=File:Dendrarium_Sochi_Mauritanian_arbour.jpg *License*: unknown *Contributors*: Ghirlandajo, Ori~, Л.П. Джепко

**File:504px-Pinguiculagrandifloralweb.jpg** *Source*: http://en.wikipedia.org/w/index.php?title=File:504px-Pinguiculagrandifloralweb.jpg *License*: unknown *Contributors*: Denis Barthel, Eugene van der Pijll, Guérin Nicolas, Michau Sm, Nilfanion, Ollin, Orchi, Taketa, 3 anonymous edits

**File:Hydrogen Density Plots.png** *Source*: http://en.wikipedia.org/w/index.php?title=File:Hydrogen_Density_Plots.png *License*: unknown *Contributors*: PoorLeno (talk)

**File:Planets2008.jpg** *Source*: http://en.wikipedia.org/w/index.php?title=File:Planets2008.jpg *License*: unknown *Contributors*: User:Farry

**File:NGC 4414 (NASA-med).jpg** *Source*: http://en.wikipedia.org/w/index.php?title=File:NGC_4414_(NASA-med).jpg *License*: unknown *Contributors*: NASA Headquarters - Greatest Images of NASA (NASA-HQ-GRIN)

**Image:Ptolemaicsystem-small.png** *Source*: http://en.wikipedia.org/w/index.php?title=File:Ptolemaicsystem-small.png *License*: unknown *Contributors*: User:Fastfission

**Image:Hubble ultra deep field high rez edit1.jpg** *Source*: http://en.wikipedia.org/w/index.php?title=File:Hubble_ultra_deep_field_high_rez_edit1.jpg *License*: unknown *Contributors*: User:Noodle snacks

**Image:Flammarion.jpg** *Source*: http://en.wikipedia.org/w/index.php?title=File:Flammarion.jpg *License*: unknown *Contributors*: Dbachmann, Diego pmc, Kersti Nebelsiek, Kugland, Leinad-Z, Martin H., Mladifilozof, Mu, Pieter Kuiper, Ragesoss, Tales23, W!B:, Warburg, 8 anonymous edits

**Image:WMAP 2003.png** *Source*: http://en.wikipedia.org/w/index.php?title=File:WMAP_2003.png *License*: unknown *Contributors*: Original uploader was Chrisbrl88 at en.wikipedia Later version(s) were uploaded by Khailarkin, Papa November at en.wikipedia.

**File:HubbleUltraDeepFieldwithScaleComparison.jpg** *Source*: http://en.wikipedia.org/w/index.php?title=File:HubbleUltraDeepFieldwithScaleComparison.jpg *License*: unknown *Contributors*: User:Autonova

**File:Cosmological Composition - Pie Chart.png** *Source*: http://en.wikipedia.org/w/index.php?title=File:Cosmological_Composition_-_Pie_Chart.png *License*: unknown *Contributors*: User:Azcolvin429

**File:Elementary particle interactions.svg** *Source*: http://en.wikipedia.org/w/index.php?title=File:Elementary_particle_interactions.svg *License*: unknown *Contributors*: User:Stannered

**Image:Aristarchus working.jpg** *Source*: http://en.wikipedia.org/w/index.php?title=File:Aristarchus_working.jpg *License*: unknown *Contributors*: Herbythyme, Jastrow, Konstable, Leinad-Z, Rocket000, Shakko, 2 anonymous edits

**File:ThomasDiggesmap.JPG** *Source*: http://en.wikipedia.org/w/index.php?title=File:ThomasDiggesmap.JPG *License*: unknown *Contributors*: (1546?-1595)

**File:Libr0309.jpg** *Source*: http://en.wikipedia.org/w/index.php?title=File:Libr0309.jpg *License*: unknown *Contributors*: AndreasPraefcke, Aristeas, Fastfission, Mattes, Pieter Kuiper, Ragesoss, Rivi, Schimmelreiter

**File:Cassini-science-br.jpg** *Source*: http://en.wikipedia.org/w/index.php?title=File:Cassini-science-br.jpg *License*: unknown *Contributors*: Pieter Kuiper, Steff

**File:Only distance is real.svg** *Source*: http://en.wikipedia.org/w/index.php?title=File:Only_distance_is_real.svg *License*: unknown *Contributors*: User:WillowW

**File:CMB Timeline300 no WMAP.jpg** *Source*: http://en.wikipedia.org/w/index.php?title=File:CMB_Timeline300_no_WMAP.jpg *License*: unknown *Contributors*: Kaldari, 1 anonymous edits

**File:Primordial nucleosynthesis.svg** *Source*: http://en.wikipedia.org/w/index.php?title=File:Primordial_nucleosynthesis.svg *License*: unknown *Contributors*: User:Lokal_Profil

**File:Multiverse - level II.svg** *Source*: http://en.wikipedia.org/w/index.php?title=File:Multiverse_-_level_II.svg *License*: unknown *Contributors*: User:K1234567890y, User:Lokal_Profil

**Image:Human artifacts at CMNH - 37.JPG** *Source*: http://en.wikipedia.org/w/index.php?title=File:Human_artifacts_at_CMNH_-_37.JPG *License*: unknown *Contributors*: User:Piotrus

**Image:Venus of Arles Louvre Ma439 n01.jpg** *Source*: http://en.wikipedia.org/w/index.php?title=File:Venus_of_Arles_Louvre_Ma439_n01.jpg *License*: unknown *Contributors*: User:Jastrow

**Image:Isis Musei Capitolini MC744.jpg** *Source*: http://en.wikipedia.org/w/index.php?title=File:Isis_Musei_Capitolini_MC744.jpg *License*: unknown *Contributors*: User:Jastrow

**File:Akropolis-detail.jpg** *Source*: http://en.wikipedia.org/w/index.php?title=File:Akropolis-detail.jpg *License*: unknown *Contributors*: Aristeas, ChristosV, Faios Nanos, Odysses, Ronaldino

**File:Parthenon.at.Nashville.Tenenssee.01.jpg** *Source*: http://en.wikipedia.org/w/index.php?title=File:Parthenon.at.Nashville.Tenenssee.01.jpg *License*: unknown *Contributors*: Bibi Saint-Pol, Davepape, Jastrow, Kaldari, Longbow4u, Malo, Para

**File:Hypatia_Raphael_Sanzio_detail.jpg** *Source*: http://en.wikipedia.org/w/index.php?title=File:Hypatia_Raphael_Sanzio_detail.jpg *License*: unknown *Contributors*: Raffaello Santi (Wikipedia)

**File:Romuvan_ceremony_(1).PNG** *Source*: http://en.wikipedia.org/w/index.php?title=File:Romuvan_ceremony_(1).PNG *License*: unknown *Contributors*: Ycco

**File:Paganavebury.jpg** *Source*: http://en.wikipedia.org/w/index.php?title=File:Paganavebury.jpg *License*: unknown *Contributors*: user:solar

**File:Amulet Thor's hammer (copy of find from Skåne) 2010-07-10.jpg** *Source*: http://en.wikipedia.org/w/index.php?title=File:Amulet_Thor's_hammer_(copy_of_find_from_Skåne)_2010-07-10.jpg *License*: unknown *Contributors*: User:Achird

**File:Panthéisme.JPG** *Source*: http://en.wikipedia.org/w/index.php?title=File:Panthéisme.JPG *License*: unknown *Contributors*: The world Pantheist movement

**File:wikisource-logo.svg** *Source*: http://en.wikipedia.org/w/index.php?title=File:Wikisource-logo.svg *License*: unknown *Contributors*: Nicholas Moreau

**File:Edward Burnett Tylor.jpg** *Source*: http://en.wikipedia.org/w/index.php?title=File:Edward_Burnett_Tylor.jpg *License*: unknown *Contributors*: Beao, MSchnitzler2000, Solon

**File:Natural-world-title.jpg** *Source*: http://en.wikipedia.org/w/index.php?title=File:Natural-world-title.jpg *License*: unknown *Contributors*: Baguala, BaldBoris, Mepolypse

**Image:BBC-NWCollection-DVD.jpg** *Source*: http://en.wikipedia.org/w/index.php?title=File:BBC-NWCollection-DVD.jpg *License*: unknown *Contributors*: Baguala

**Image:yggdrasil.jpg** *Source*: http://en.wikipedia.org/w/index.php?title=File:Yggdrasil.jpg *License*: unknown *Contributors*: AnonMoos, Baisemain, Damouns, Gryffindor, Helios, Kjetil r, Nillerdk, 1 anonymous edits

**File:Sefan and tree.jpg** *Source*: http://en.wikipedia.org/w/index.php?title=File:Sefan_and_tree.jpg *License*: unknown *Contributors*: anonimus

**Image:AinuBearSacrificeCirca1870.jpg** *Source*: http://en.wikipedia.org/w/index.php?title=File:AinuBearSacrificeCirca1870.jpg *License*: unknown *Contributors*: User:PHGCOM, User:PHGCOM

**Image:CowHA.jpg** *Source*: http://en.wikipedia.org/w/index.php?title=File:CowHA.jpg *License*: unknown *Contributors*: User:Himalayan_Academy_Publications#Copyright_permissions

**File:Mosaic Pan Genazzano Massimo.jpg** *Source*: http://en.wikipedia.org/w/index.php?title=File:Mosaic_Pan_Genazzano_Massimo.jpg *License*: unknown *Contributors*: User:Jastrow

**Image:Dog worship in Hinduism.jpg** *Source*: http://en.wikipedia.org/w/index.php?title=File:Dog_worship_in_Hinduism.jpg *License*: unknown *Contributors*: User:Wiki-uk

**Image:13th century Ganesha statue.jpg** *Source*: http://en.wikipedia.org/w/index.php?title=File:13th_century_Ganesha_statue.jpg *License*: unknown *Contributors*: G.dallorto, Geofrog, Gildemax, Jastrow, Jat, Johnbod, Paddy, Quadell, Redtigerxyz, Roland zh, Wst, 5 anonymous edits

**Image:Horus standing.svg** *Source*: http://en.wikipedia.org/w/index.php?title=File:Horus_standing.svg *License*: unknown *Contributors*: User:Jeff Dahl

**Image:Naga3.jpg** *Source*: http://en.wikipedia.org/w/index.php?title=File:Naga3.jpg *License*: unknown *Contributors*: Pratheepps at en.wikipedia

**Image:Quetzalcoatl telleriano.jpg** *Source*: http://en.wikipedia.org/w/index.php?title=File:Quetzalcoatl_telleriano.jpg *License*: unknown *Contributors*: Kilom691, Mazan, Onderwijsgek, Ptcamn

**Image:Dagon1.jpg** *Source*: http://en.wikipedia.org/w/index.php?title=File:Dagon1.jpg *License*: unknown *Contributors*: Original uploader was MT-DOGMA at en.wikipedia

**File:Greece Mount Olympus (1).jpg** *Source*: http://en.wikipedia.org/w/index.php?title=File:Greece_Mount_Olympus_(1).jpg *License*: unknown *Contributors*: Alina Zienowicz Ala z

**File:Machapuchre-wow.jpg** *Source*: http://en.wikipedia.org/w/index.php?title=File:Machapuchre-wow.jpg *License*: unknown *Contributors*: Nancy Collins

MIX
Papier aus verantwortungsvollen Quellen
Paper from responsible sources
FSC® C105338

Printed by Books on Demand GmbH, Norderstedt / Germany